Bombay Lectures on
HIGHEST WEIGHT REPRESENTATIONS
—————————— *of* ——————————
INFINITE DIMENSIONAL LIE ALGEBRAS

Second Edition

ADVANCED SERIES IN MATHEMATICAL PHYSICS

Editors-in-Charge

H Araki (*RIMS, Kyoto*)
V G Kac (*MIT*)
D H Phong (*Columbia University*)

Associate Editors

L Alvarez-Gaumé (*CERN*)
J P Bourguignon (*Ecole Polytechnique, Palaiseau*)
T Eguchi (*University of Tokyo*)
B Julia (*CNRS, Paris*)
F Wilczek (*MIT*)

*Published**

*For the complete list of volumes in this series, please visit
www.worldscientific.com/series/asmp

Advanced Series in Mathematical Physics – Vol. 29

Bombay Lectures on

HIGHEST WEIGHT REPRESENTATIONS

of

INFINITE DIMENSIONAL LIE ALGEBRAS

Second Edition

Victor G. Kac
Massachusetts Institute of Technology, USA

Ashok K. Raina
Tata Institute of Fundamental Research, India

Natasha Rozhkovskaya
Kansas State University, USA

 **World Scientific**

NEW JERSEY · LONDON · SINGAPORE · BEIJING · SHANGHAI · HONG KONG · TAIPEI · CHENNAI

Published by

World Scientific Publishing Co. Pte. Ltd.
5 Toh Tuck Link, Singapore 596224
USA office: 27 Warren Street, Suite 401-402, Hackensack, NJ 07601
UK office: 57 Shelton Street, Covent Garden, London WC2H 9HE

British Library Cataloguing-in-Publication Data
A catalogue record for this book is available from the British Library.

Cover image: Lie algebra tree, oil on canvas, by Natasha Koshka, from the art collection of Harvard Department of Mathematics; reproduced with the permission of the Harvard Department of Mathematics.

Advanced Series in Mathematical Physics — Vol. 29
BOMBAY LECTURES ON HIGHEST WEIGHT REPRESENTATIONS OF INFINITE DIMENSIONAL LIE ALGEBRAS
(Second Edition)

ISBN 978-981-4522-18-2
ISBN 978-981-4522-19-9 (pbk)

Printed in Singapore

PREFACE

This book is a write-up of a series of lectures given by the first author at the Tata Institute, Bombay, during December 1985–January 1986.

The dominant theme of these lectures is the idea of a highest weight representation. This idea goes through four different incarnations.

The first is the canonical commutation relations of the infinite-dimensional Heisenberg algebra (= oscillator algebra). Although this example is extremely simple, it not only contains the germs of the main features of the theory, but also serves as a basis for most of the constructions of representations of infinite-dimensional Lie algebras.

The second is the highest weight representations of the Lie algebra $g\ell_\infty$ of infinite matrices, along with their applications to the theory of soliton equations, discovered by Sato and Date-Jimbo-Kashiwara-Miwa. Here the main point is the isomorphism between the vertex and the "Dirac sea" realizations of the fundamental representations of $g\ell_\infty$, a kind of a Bose-Fermi correspondence.

The third is the unitary highest weight representations of the affine Kac-Moody (= current) algebras. Since there is now a book devoted to the theory of Kac-Moody algebras, it was decided to devote to them a minimum attention. In the lectures affine algebras play a prominent role only in the Sugawara construction as the main tool in the study of the fourth incarnation of the main idea, the theory of highest weight representations of the Virasoro algebra.

The main results of the representation theory of the Virasoro algebra which are proved in these lectures are the Kac determinant formula and the unitarity of the "discrete series" representations of Belavin-Polyakov-Zamolodchikov and Friedan-Qiu-Shenker.

We hope that this elementary introduction to the subject, written by a mathematician and a physicist, will prove useful to both mathematicians and physicists. To mathematicians, since it illustrates, on important examples, the interaction of the key ideas of the representation theory of infinite-dimensional Lie algebras; and to physicists, since this theory is turning before our very eyes into an important component of such domains of theoretical physics as soliton theory, theory of two-dimensional statistical models, and string theory.

Throughout the book, the base field is the field of complex numbers $\mathbb{C}$, unless otherwise stated, $\mathbb{R}$ denotes the set of real numbers, $\mathbb{Z}$ the set of integers, and $\mathbb{Z}_+$ (resp. $\mathbb{N}$) the set of nonnegative (resp. positive) integers.

The authors wish to thank the participants of the lectures, especially S. M. Roy and S. R. Wadia, for valuable suggestions and comments.

PREFACE TO THE SECOND EDITION

The first edition of this book was based on the series of lectures, given by the first author at the Tata Institute of Fundamental Research, Mumbai (Bombay), in the winter of 1985–1986, just before the birth of the theory of vertex algebras.

As a consequence of the development of this theory, many results and constructions of representation theory of infinite-dimensional Lie algebras have been greatly extended and clarified. Therefore, when the first author was again invited to visit the TIFR some eighteen years later, it was quite natural to lecture on vertex algebras.

The second edition of the book consists of two parts. The first part (Lectures 1-12) contains the largely unchanged text of the first edition, while the second part (Lectures 13-18) is an extended write-up of the lectures, delivered at the TIFR in January 2003.

The basic idea of these lectures was to demonstrate how the key notions of the theory of vertex algebras - such as the quantum fields, their normal ordered product and λ-bracket, energy-momentum field and conformal weight, untwisted and twisted representations - simplify and clarify the constructions of the first edition of the book.

The first author wishes to thank M. Shubin for a correction to Proposition 1.2, and B. Bakalov for a discussion on his approach to twisted representations, used in the book.

CONTENTS

LECTURE 1

In this series of lectures we shall demonstrate some basic concepts and methods of the representation theory of infinite-dimensional Lie algebras on four main examples:

1. The oscillator algebra
2. The Virasoro algebra
3. The Lie algebra $g\ell_\infty$
4. The affine Kac-Moody algebras.

In fact, the interplay between these examples is one of the key methods of the theory.

1.1. The Lie algebra $\mathfrak{d}$ of complex vector fields on the circle

We shall first consider the Virasoro algebra, which is playing an increasingly important role in theoretical physics. It is also a natural algebra to consider from a mathematical point of view as it is a central extension of the complexification of the Lie algebra $Vect\ S^1$ of (real) vector fields on the circle S^1. We shall start by finding the structure of $Vect\ S^1$ and later consider its central extensions.

Any element of $Vect\ S^1$ is of the form $f(\theta)\ d/d\theta$, where $f(\theta)$ is a smooth real-valued function on S^1, with θ a parameter on S^1 and $f(\theta+2\pi) = f(\theta)$. The Lie bracket of vector fields is:

$$\left[f(\theta)\frac{d}{d\theta}, g(\theta)\frac{d}{d\theta}\right] = (fg' - f'g)(\theta)\frac{d}{d\theta},$$

where prime stands for the derivative. A basis (over $\mathbb{R}$) for $Vect\ S^1$ is

1

provided by the vector fields

$$\frac{d}{d\theta}, \quad \cos(n\theta)\frac{d}{d\theta}, \quad \sin(n\theta)\frac{d}{d\theta} \quad (n = 1, 2, \ldots).$$

To avoid convergence questions we consider this as a vector space basis, so that $f(\theta)$, $g(\theta)$ are arbitrary trigonometric polynomials, and take its linear span over $\mathbb{C}$ as this permits us to introduce $\exp(in\theta)$ instead of $\cos(n\theta)$ and $\sin(n\theta)$. We thus obtain a complex Lie algebra, denoted by $\mathfrak{d}$, with a basis

$$d_n = i\exp(in\theta)\frac{d}{d\theta} = -z^{n+1}\frac{d}{dz} \quad (n \in \mathbb{Z}) \tag{1.1}$$

where $z = \exp(i\theta)$. These elements satisfy the following commutation relations:

$$[d_m, d_n] = (m - n)d_{m+n} \quad (m, n \in \mathbb{Z}). \tag{1.2}$$

The Lie algebra *Vect* S^1 can be considered as the Lie algebra of the group G of orientation preserving diffeomorphisms of S^1. If ζ_1, ζ_2 are two elements of G then their product is defined by composition:

$$(\zeta_1 \cdot \zeta_2)(z) = \zeta_1(\zeta_2(z))$$

for each $z = \exp(i\theta)$ on S^1. If $f(z)$ is an element of the vector space of smooth complex-valued functions on S_1, then $\gamma \in G$ acts on $f(z)$ by

$$\pi(\gamma)f(z) = f(\gamma^{-1}(z)). \tag{1.3}$$

This clearly defines a representation of G. We take γ close to the identity (as physicists do):

$$\gamma(z) = z(1 + \epsilon(z)) = z + \sum_{n=-\infty}^{\infty} \epsilon_n z^{n+1} \tag{1.4a}$$

where we have made a Laurent (or Fourier) expansion of $\epsilon(z)$ and the ϵ_n are to be retained up to first order only. Then

$$\gamma^{-1}(z) = z - \sum_{n=-\infty}^{\infty} \epsilon_n z^{n+1} \tag{1.4b}$$

and

$$\pi(\gamma)f(z) = f\left(z - \sum_{n} \epsilon_n z^{n+1}\right) = \left(1 + \sum_{n} \epsilon_n d_n\right)f(z) \tag{1.5}$$

where the d_n are defined by (1.1). This shows that the d_n form a (topological) basis of the complexification of the Lie algebra of G.

In the following we shall consider the complex Lie algebra $\mathfrak{d}$ and view $\mathfrak{d} \cap Vect\, S^1$ as the subalgebra (over $\mathbb{R}$) of real elements. One way to do this is to regard $\mathfrak{d} \cap Vect\, S^1$ as the subalgebra of fixed points for the operation of complex conjugation under which d_n maps to $-d_{-n}$ and a scalar λ to its complex conjugate $\overline{\lambda}$. It is more convenient, however, to introduce a slightly different operation defined by:

$$\omega(d_n) = d_{-n}, \tag{1.6a}$$

$$\omega(\lambda x) = \overline{\lambda}\omega(x), \tag{1.6b}$$

so that

$$\omega([x,y]) = [\omega(y), \omega(x)] \tag{1.6c}$$

where $x, y \in \mathfrak{d}$, $\lambda \in \mathbb{C}$. Thus ω is an antilinear anti-involution having the algebraic properties of Hermitian conjugation. Now $\mathfrak{d} \cap Vect\, S^1$ consists of elements of $\mathfrak{d}$ fixed under $-\omega$.

The purely algebraic operation ω on $\mathfrak{d}$ can become an adjoint operation with respect to a suitable scalar product if we have a representation of $\mathfrak{d}$ in some vector space. Suppose that we have a unitary representation of the group G on a vector space V with a positive-definite Hermitian form $\langle \cdot \,|\, \cdot \rangle$. Identifying elements of G with corresponding operators, we have:

$$\langle g(u)\,|\,g(v)\rangle = \langle u\,|\,v\rangle \quad \text{for} \ \ g \in G, \ u, v \in V.$$

Going over to the Lie algebra, this means that

$$\langle x(u)\,|\,v\rangle = -\langle u\,|\,x(v)\rangle \quad \text{for} \ \ x \in Vect\, S^1,$$

and for any $x \in \mathfrak{d}$:

$$\langle x(u)\,|\,v\rangle = \langle u\,|\,\omega(x)(v)\rangle. \tag{1.7}$$

This motivates the following definitions:

Definition 1.1. Let $\mathfrak{g}$ be a Lie algebra and let ω be an antilinear anti-involution on $\mathfrak{g}$, i.e. an $\mathbb{R}$-linear involution satisfying (1.6b and c). Let V be a representation space of $\mathfrak{g}$ and $\langle \cdot \,|\, \cdot \rangle$ an Hermitian form on V. We say that $\langle \cdot | \cdot \rangle$ is *contravariant* if (1.7) holds for all $x \in \mathfrak{g}$, and u, $v \in V$. When $\langle \cdot | \cdot \rangle$ is nondegenerate, this means that

$$x^\dagger = \omega(x) \quad \text{for all} \quad x \in \mathfrak{g}. \tag{1.7'}$$

Here and further $x^\dagger$ stands for the Hermitian adjoint of the operator x. We further say that this representation is *unitary* if in addition

$$\langle v | v \rangle > 0 \quad \text{for all} \quad v \in V, \quad v \neq 0.$$

1.2. Representations $V_{\alpha,\beta}$ of $\mathfrak{d}$

We shall find representations of $\mathfrak{d}$ by considering a suitable vector space on which the group G acts and determining the action of G in this space for elements close to the identity. Using (1.5) we shall determine the action of d_n in this vector space.

Let $V_{\alpha,\beta}$ denote that space of 'densities' of the form $P(z) z^\alpha (dz)^\beta$, where α and β are complex numbers and $P(z)$ is an arbitrary polynomial in z and z^{-1}. A basis for $V_{\alpha,\beta}$ is given by the set of vectors

$$v_k = z^{k+\alpha} (dz)^\beta \quad (k \in \mathbb{Z}). \tag{1.8}$$

From (1.3),

$$\pi(\gamma) v_k = (\gamma^{-1}(z))^{k+\alpha} (d\gamma^{-1}(z))^\beta$$

and if γ is of the form (1.4a), we can use (1.4b) for $\gamma^{-1}(z)$. Thus

$$\pi(\gamma) v_k = \left(z - \sum_n \epsilon_n z^{n+1} \right)^{k+\alpha} \left(\left(1 - \sum_n \epsilon_n(n+1)z^n \right) dz \right)^\beta$$

$$= \left(1 - (k+\alpha) \sum_n \epsilon_n z^n \right) \left(1 - \beta \sum_n \epsilon_n(n+1)z^n \right) z^{k+\alpha} (dz)^\beta$$

$$= \left(1 - \sum_n \epsilon_n(k+\alpha+\beta n+\beta)z^n \right) z^{k+\alpha} (dz)^\beta.$$

Comparing with (1.5) we see that

$$d_n(v_k) = -(k+\alpha+\beta+\beta n)v_{n+k} \quad (n, k \in \mathbb{Z}). \tag{1.9}$$

Formula (1.9) defines a two-parameter family of representations of the Lie algebra $\mathfrak{d}$. Note from (1.9) that d_0 is diagonal:

$$d_0(v_k) = -(k + \alpha + \beta)v_k. \tag{1.10}$$

The operator d_0 is called the *energy operator*.

The following well-known lemma is useful in the proof of irreducibility:

Lemma 1.1. Let V be a representation of a Lie algebra $\mathfrak{g}$ which decomposes as a direct sum of eigenspaces of a finite-dimensional commutative subalgebra $\mathfrak{h}$:

$$V = \bigoplus_{\lambda \in \mathfrak{h}^*} V_\lambda \tag{1.11a}$$

where $V_\lambda = \{v \in V \mid h(v) = \lambda(h)v \text{ for all } h \in \mathfrak{h}\}$, and $\mathfrak{h}^*$ is the dual vector space of $\mathfrak{h}$. Then any subrepresentation U of V respects this decomposition in the sense that

$$U = \bigoplus_\lambda (U \cap V_\lambda). \tag{1.12a}$$

Applying this to our case with $\mathfrak{g} = \mathfrak{d}$ and $\mathfrak{h} = \mathbb{C}\,d_0$ we obtain the following corollary:

Corollary 1.1. Let V be a representation of $\mathfrak{d}$ which decomposes as a direct sum of eigenspaces of d_0:

$$V = \bigoplus_k V_k. \tag{1.11b}$$

Then any subrepresentation U of V respects this decomposition:

$$U = \bigoplus_k (U \cap V_k). \tag{1.12b}$$

Proof. We first prove Corollary 1.1. Any $v \in V$ can be written in the form $v = \sum_{j=1}^m w_j$, where $w_j \in V_{\lambda_j}$ according to (1.11b), and $d_0(w_j) = \lambda_j w_j$, where $\lambda_j \neq \lambda_k$ for $j \neq k$ $(j, k = 1, \ldots, m)$. Then if $v \in U$, we have the following set of equations:

$$v = w_1 + w_2 + \ldots + w_m$$
$$d_0(v) = \lambda_1 w_1 + \lambda_2 w_2 + \ldots + \lambda_m w_m$$

$$d_0^2(v) = \lambda_1^2 w_1 + \lambda_2^2 w_2 + \ldots + \lambda_m^2 w_m$$

$$\cdots\cdots\cdots\cdots\cdots\cdots\cdots\cdots\cdots\cdots\cdots\cdots\cdots$$

$$d_0^{m-1}(v) = \lambda_1^{m-1} w_1 + \lambda_2^{m-1} w_2 + \ldots + \lambda_m^{m-1} w_m.$$

Since U is a subrepresentation of $\mathfrak{d}$, it follows that $v, d_0(v), \ldots, d_0^{m-1}(v)$ must all lie in U. We then have a system of equations with a matrix which is invertible since its determinant is a Vandermonde and hence not zero. Thus each of the w_j $(j = 1, \ldots, m)$ can be expressed as a linear combination of vectors in U, proving (1.12b). The proof of Lemma 1.1 is identical to that of the corollary, since we can always find $h \in \mathfrak{h}$ such that $\lambda_j(h) \neq \lambda_k(h)$ for $j \neq k$ and so we merely have to replace d_0 by this h in the above proof. ∎

Remark 1.1. The representation $V_{\alpha,\beta}$ is isomorphic to the representation $V_{\alpha+m,\beta}$ for $m \in \mathbb{Z}$, by the shift by m of indices of the basis. ∎

Proposition 1.1. The representation $V_{\alpha,\beta}$ of $\mathfrak{d}$ is reducible if (i) $\alpha \in \mathbb{Z}$ and $\beta = 0$, or (ii) $\alpha \in \mathbb{Z}$ and $\beta = 1$; otherwise it is irreducible.

Proof. Let U be a nonzero subrepresentation of $V_{\alpha,\beta}$. Then by Corollary 1.1, U is also a direct sum of some of the 1-dimensional subspaces $\mathbb{C}v_j$. Let $v_k \in U$. If U is not the whole of $V_{\alpha,\beta}$, there is at least one vector in $V_{\alpha,\beta}\backslash U$. Let us first suppose that there are at least two distinct vectors v_m and $v_n(m \neq n)$ in $V_{\alpha,\beta}\backslash U$. By (1.9),

$$d_{m-k}(v_k) = -(k + \alpha + \beta + \beta(m - k))v_m, \qquad (1.13)$$

$$d_{n-k}(v_k) = -(k + \alpha + \beta + \beta(n - k))v_n. \qquad (1.14)$$

Since U is a representation of $\mathfrak{d}$, the right-hand sides of (1.13) and (1.14) must lie in U, which is only possible if

$$k + \alpha + \beta + \beta(m - k) = 0 = k + \alpha + \beta + \beta(n - k).$$

Since $m \neq n$ we have $\beta = 0$ and so $k + \alpha = 0$. This shows that if U has codimension at least 2, then $\beta = 0$ and $\alpha \in \mathbb{Z}$. We may assume, from Remark 1.1, that $\alpha = 0$ so that $U = \mathbb{C}v_0$. Thus $V_{0,0}' = V_{0,0}/\mathbb{C}v_0$ is an irreducible representation of $\mathfrak{d}$.

Let us now suppose that U has codimension 1 and $v_m \in V_{\alpha,\beta}\backslash U$. Hence U has at least two vectors v_k, v_ℓ $(k \neq \ell \neq m)$. Thus $d_{m-k}v_k$ and $d_{m-\ell}v_\ell$

lie in U, which implies by (1.9) that

$$k + \alpha + \beta(m - k + 1) = 0 = \ell + \alpha + \beta(m - \ell + 1).$$

We thus get $\beta = 1$ and from Remark 1.1 we may assume that $\alpha = 0$, so that $m = -1$ and $U = \{\text{linear span of } v_k \mid k \in \mathbb{Z}, \, k \neq -1\}$. Thus, $V'_{0,1} = U$ is an irreducible representation of $\mathfrak{d}$. $\blacksquare$

We put $V'_{\alpha,\beta} = V_{\alpha,\beta}$ if $V_{\alpha,\beta}$ is irreducible; otherwise, $V'_{\alpha,\beta}$ is as in the proof of Proposition 1.1, so that all $V'_{\alpha,\beta}$ are irreducible.

Only some of the irreducible representations $V'_{\alpha,\beta}$ can have a Hermitian contravariant form:

Proposition 1.2. Representation $V'_{\alpha,\beta}$ has a nondegenerate Hermitian contravariant form if and only if (1) $\beta + \bar{\beta} = 1$ and $\alpha + \beta \in \mathbb{R}$, or (2) $\beta = 0$ or $\beta = 1$ and $\alpha \in \mathbb{R}$. The form is unitary in the case (1) and non-unitary in the case (2).

Proof. We leave the straightforward proof to the reader. $\blacksquare$

1.3. Central extensions of $\mathfrak{d}$: the Virasoro algebra

We shall now study Lie algebra extensions $\hat{\mathfrak{d}}$ of $\mathfrak{d}$ by a 1-dimensional center $\mathbb{C}C$. This means that

$$\hat{\mathfrak{d}} = \mathfrak{d} \oplus \mathbb{C}C$$

and the commutation relations (1.2) are replaced by commutation relations of the form

$$
\begin{aligned}
[d_m, d_n] &= (m - n)d_{m+n} + a(m,n)C, \\
[d_m, C] &= 0,
\end{aligned}
\tag{1.15}
$$

where $a(m,n) \in \mathbb{C}$ and $m, n \in \mathbb{Z}$.

The function $a(m,n)$ cannot be arbitrary because of the anticommutativity of the bracket and the Jacobi identity.

We observe from (1.15) that putting $d'_0 = d_0$, $d'_n = d_n - [a(0,n)/n]C$ ($n \neq 0$) we have $[d'_0, d'_n] = -nd'_n$ ($n \in \mathbb{Z}$). This transformation is merely a change of basis in $\hat{\mathfrak{d}}$ and so we can drop the primes and say that

$$[d_0, d_n] = -nd_n \ (n \in \mathbb{Z}).\tag{1.16}$$

From the Jacobi identity for d_0, d_m, d_n we get

$$[d_0, [d_m, d_n]] = -(m + n)[d_m, d_n].\tag{1.17}$$

Substituting (1.15) in (1.17) and using (1.16) we get $(m+n)a(m,n)C = 0$. This shows that $a(m, n) = \delta_{m,-n}a(m)$, so that (1.15) becomes:

$$[d_m, d_n] = (m - n)d_{m+n} + \delta_{m,-n}a(m)C \quad (m, n \in \mathbb{Z})\tag{1.18}$$

where $a(m) = -a(-m)$ by anticommutativity. We now work out the Jacobi identity for d_ℓ, d_m, d_n with $\ell+m+n = 0$ using (1.18) and the antisymmetry of $a(m)$. We get

$$(m - n)a(m + n) - (2n + m)a(m) + (n + 2m)a(n) = 0.\tag{1.19}$$

Putting $n = 1$ in (1.19) we get

$$(m - 1)a(m + 1) = (m + 2)a(m) - (2m + 1)a(1).\tag{1.20}$$

Since $a(-m) = -a(m)$ we have $a(0) = 0$ and we have to solve (1.20) for positive values of m only. Equation (1.20) is a linear recursion relation and its space of solutions is at most 2-dimensional, since the knowledge of $a(1)$ and $a(2)$ gives all $a(m)$. We observe that $a(m) = m$ and $a(m) = m^3$ are solutions. Hence $a(m) = \alpha m + \beta m^3$ is the general solution. If $\beta = 0$ then by defining $d_0' = d_0 + 1/2\alpha C$, $d_i' = d_i \ (i \neq 0)$, the $d_i' \ (i \in \mathbb{Z})$ span an algebra without central charge, i.e. $\mathfrak{d}$, so that $\hat{\mathfrak{d}}$ is a direct sum of Lie algebras $\mathfrak{d}$ and $\mathbb{C}C$. Hence, for a nontrivial central extension, $\beta \neq 0$ while α can be chosen arbitrarily; we conventionally choose $\alpha = -\beta$ so that $a(m) = \beta(m^3 - m)$. By rescaling C, we can choose a fixed value for β. Conventionally we put $\beta = 1/12$. We thus arrive at the *Virasoro algebra*, the Lie algebra *Vir* with a basis

$$\{d_m, m \in \mathbb{Z}; C\}$$

and the following commutation relations

$$[d_m, C] = 0,\tag{1.21a}$$

$$[d_m, d_n] = (m - n)d_{m+n} + \delta_{m,-n}\frac{(m^3 - m)}{12}C.\tag{1.21b}$$

Thus, we have proved the following

Proposition 1.3. Every nontrivial central extension of the Lie algebra $\mathfrak{d}$ by a 1-dimensional center is isomorphic to the Virasoro algebra *Vir*. ∎
We have obtained along with the proof:

Corollary 1.2. If $[d_m, d_n] = (m - n)d_{m+n} + \delta_{m,-n}a(m)C$ defines a Lie algebra, then $a(m) = \alpha m + \beta m^3$ for some $\alpha, \beta \in \mathbb{C}$. ∎

In these lectures we shall be mostly concerned with the unitary representations of *Vir*. Unitarity is defined through a Hermitian contravariant form (Definition 1.1) and contravariance is defined in terms of an antilinear, anti-involution ω of *Vir* defined by (cf. (1.6)):

$$\omega(d_n) = d_{-n} \quad (n \in \mathbb{Z}), \tag{1.22a}$$

$$\omega(C) = C. \tag{1.22b}$$

In particular, d_0 and C are self-adjoint elements of *Vir*.

LECTURE 2

2.1. Definition of positive-energy representations of *Vir*

In the previous lecture we introduced the Virasoro algebra

$$Vir = \mathbb{C}C + \sum_{n \in \mathbb{Z}} \mathbb{C}d_n \,,$$

$$[C, d_n] = 0 \,, \tag{2.1}$$

$$[d_m, d_n] = (m - n)d_{m+n} + \delta_{m,-n} \frac{m^3 - m}{12} C \,,$$

as a central extension of $\mathfrak{d}$, the Lie algebra of complex polynomial vector fields on S^1. The so-called "two-cocycle" $a(m, n) = \delta_{m,-n} \, (m^3 - m)/12$ was probably first discovered by Gelfand and Fuchs [1968] who computed the entire cohomology ring of $\mathfrak{d}$. They showed that the algebra $H^*(\mathfrak{d})$ is the tensor product of the algebra of polynomials generated by a single generator of degree 2 and an exterior algebra defined by a single generator of degree 3, and hence

$$H^j(\mathfrak{d}) = \mathbb{C} \text{ for } j = 0, 2, 3, \dots, H^1(\mathfrak{d}) = 0 \,.$$

What we proved in the last lecture is actually that $H^2(\mathfrak{d}) = \mathbb{C}$. (The equality $H^1(\mathfrak{d}) = 0$ simply means that $[\mathfrak{d}, \mathfrak{d}] = \mathfrak{d}$.)

In the previous lecture we also constructed some representations of *Vir* with $C = 0$. As we can see from (1.10), in these representations the spectrum of d_0 is unbounded both from above and from below. We shall continue to use the terminology "energy operator" for d_0 as an element of *Vir*. This leads us naturally to define a 'positive-energy representation' of *Vir*:

Definition 2.1. A representation of Vir is called a *positive-energy* representation if d_0 is diagonal and all its eigenvalues are nonnegative.

We shall see that Vir has nontrivial positive-energy unitary representations only if $C \neq 0$ and this is one of the reasons why Vir is so much more interesting an algebra than $\mathfrak{d}$. We shall learn in Lecture 4 how to find positive-energy representations of Vir using Dirac's hole theory. Now, however, we consider the Virasoro [1970] construction of some unitary positive-energy representations of Vir in terms of bosonic oscillators.

2.2. Oscillator algebra $\mathscr{A}$

Let $\mathscr{A}$ be the *oscillator* (Heisenberg) algebra, the complex Lie algebra with a basis $\{a_n, n \in \mathbb{Z}; \mathbb{1}\}$, and the commutation relations

$$[\mathbb{1}, a_n] = 0 \ (n \in \mathbb{Z}) \,,$$
$$[a_m, a_n] = m\delta_{m,-n}\mathbb{1} \quad (m, n \in \mathbb{Z}) \,. \tag{2.2}$$

We note that $[a_0, a_n] = 0 \ (n \in \mathbb{Z})$ so that a_0 is a central element (zero mode). We use the notation $\mathbb{1}$ since in all representations of $\mathscr{A}$ that we shall consider, $\mathbb{1}$ will be represented by the identity operator.

Introduce the Fock space $B = \mathbb{C}[x_1, x_2, \ldots]$; this is the space of polynomials in infinitely many variables $x_1, x_2, \ldots$.

Given $\mu \in \mathbb{R}$, define the following representation of $\mathscr{A}$ on B ($n \in \mathbb{N}$):

$$a_n = \epsilon_n \partial/\partial x_n \,, \tag{2.3a}$$

$$a_{-n} = \epsilon_n^{-1} n x_n \,, \tag{2.3b}$$

$$a_0 = \mu I \,, \tag{2.3c}$$

$$\mathbb{1} = I \,. \tag{2.3d}$$

It is clear that these operators satisfy (2.2). The ϵ_n are arbitrary real scale factors which will be useful later on. As such they are inessential, but this is not the case for the parameter μ.

Lemma 2.1. The representation (2.3) of $\mathscr{A}$ is irreducible.

Proof. Any polynomial in B can be reduced to a multiple of 1 by successive application of the a_n with $n > 0$ (the annihilation operators). Then successive application of the a_{-n} with $n > 0$ (creation operators) can give any other polynomial in B. ∎

The constant polynomial $v = 1$, which is called the *vacuum vector* of B, has the properties

$$a_n(v) = 0 \quad \text{for} \ \ n > 0, \tag{2.4a}$$

$$a_0(v) = \mu v, \tag{2.4b}$$

$$\mathbb{1}(v) = v. \tag{2.4c}$$

Proposition 2.1. Let V be a representation of $\mathscr{A}$ which admits a nonzero vector v satisfying (2.4). Then monomials of the form $a_{-1}^{k_1} \ldots a_{-n}^{k_n}(v)$ ($k_i \in \mathbb{Z}_+$) are linearly independent. If these monomials span V, then V is equivalent to the representation of $\mathscr{A}$ on B given by (2.3). In particular, this is the case if V is irreducible.

Proof. We have a mapping ϕ from B to V defined by $\phi(P(\ldots, x_n, \ldots)) = P(\ldots, (\epsilon_n/n)a_{-n}, \ldots)v$. It is clear that if P is an element of B, then $a_n(\phi(P)) = \phi(a_n(P))$, i.e. ϕ is an intertwining operator. Since B is irreducible, $\ker \phi = 0$ and so ϕ is an isomorphism if ϕ is onto. ∎

We define an antilinear anti-involution ω on $\mathscr{A}$ by

$$\omega(a_n) = a_{-n}, \ \ \omega(\mathbb{1}) = \mathbb{1},$$

and extend it to an antilinear anti-involution of the universal enveloping algebra $U(\mathscr{A})$ of $\mathscr{A}$. (Physicists use the notation $a_n^{\dagger} = a_{-n}$ instead.)

Proposition 2.2. Let V be as in Proposition 2.1. Then V carries a unique Hermitian form $\langle \cdot \, | \, \cdot \rangle$ which is contravariant with respect to ω, and such that $\langle v \, | \, v \rangle = 1$ for the vacuum vector v. The distinct monomials $a_{-1}^{k_1} \ldots a_{-n}^{k_n}(v)$ ($k_i \in \mathbb{Z}_+$) form an orthogonal basis with respect to $\langle \cdot \, | \, \cdot \rangle$. These monomials have norms given by

$$\langle a_{-1}^{k_1} \ldots a_{-n}^{k_n} v \, | \, a_{-1}^{k_1} \ldots a_{-n}^{k_n} v \rangle = \prod_{j=1}^{n} k_j! j^{k_j}. \tag{2.5}$$

Proof. If $\langle \cdot \, | \, \cdot \rangle$ is a contravariant Hermitian form, then both the orthogonality and (2.5) are proved by induction on $k_1 + \ldots + k_n$, proving uniqueness.

One checks directly that the Hermitian form, for which monomials are orthogonal and have norms given by (2.5), is contravariant, proving existence. (A more conceptual proof of existence is given below.) ∎

Corollary 2.1. The contravariant Hermitian form on V such that $\langle v \, | \, v \rangle = 1$ is positive-definite. ∎

Definition 2.2. Let P be an arbitrary polynomial in B. The *vacuum expectation value* of P, denoted by $\langle P \rangle$, is defined as the constant term of P.

One clearly has the following useful formula:

$$\langle \omega(a)(1) \rangle = \overline{\langle a(1) \rangle} \quad \text{for} \quad a \in U(\mathscr{A}), \tag{2.6a}$$

where $U(\mathscr{A})$ is the universal enveloping algebra of $\mathscr{A}$. We can now define for $a, b \in U(\mathscr{A})$:

$$\langle a(1) \, | \, b(1) \rangle = \langle \omega(a)b(1) \rangle . \tag{2.6b}$$

One checks using (2.6a) that this is a Hermitian form; it is obviously contravariant and $\langle 1 \, | \, 1 \rangle = 1$. Hence, by Proposition 2.2, formulas (2.5) and (2.6b) are equivalent.

Definition 2.3. Define the degree of the monomial $x_1^{j_1} \ldots x_k^{j_k}$ as $j_1 + 2j_2 + \ldots + kj_k$. Let B_j be the subspace of B spanned by monomials of degree j. B_j is clearly finite dimensional and dim $B_j = p(j)$. Here and further $p(j)$ denotes the number of partitions of $j \in \mathbb{Z}_+$ into a sum of positive integers with $p(0) = 1$. We have

$$B = \bigoplus_{j \geqslant 0} B_j , \tag{2.7}$$

the *principal gradation* of B. We define the *q-dimension* of B to be

$$\dim_q B = \sum_{j \geqslant 0} (\dim B_j) q^j .$$

Then we clearly have:

$$\dim_q B = \sum_{j \geqslant 0} p(j)q^j = 1/\varphi(q) \tag{2.8a}$$

where

$$\varphi(q) = \prod_{j \in \mathbb{N}} (1 - q^j) . \tag{2.8b}$$

Remark 2.1. The monomial $x_1^{j_1} \ldots x_k^{j_k}$ represents j_1 oscillators in state $1, j_2$ in state 2, etc. and hence the degree is essentially the energy of the state.

Putting $q = \exp(-\beta)$ we see that 'q-dimension' is related to the partition function of statistical mechanics. ∎

2.3. Oscillator representations of *Vir*

Our aim in introducing the oscillator algebra $\mathscr{A}$ and its Fock representation is to introduce the Virasoro operators L_k. These are defined in the Fock representation B with $a_0 = \mu$ by:

$$L_k = \frac{1}{2} \sum_{j \in \mathbb{Z}} :a_{-j}a_{j+k}: \quad (k \in \mathbb{Z}), \tag{2.9}$$

where the colons indicate 'normal ordering', defined by

$$:a_i a_j: = \begin{cases} a_i a_j & \text{if } i \leqslant j \\ a_j a_i & \text{if } i > j. \end{cases} \tag{2.10}$$

As a result of normal ordering, when the operator L_k is applied to any vector of B, only a finite number of terms in the sum contribute. Hence L_k makes sense in B. The fundamental property of the L_k is that they provide a representation of the Virasoro algebra *Vir* in B with central charge $C = 1$:

Proposition 2.3. The L_k satisfy the commutation relations

$$[L_m, L_n] = (m - n)L_{m+n} + \delta_{m,-n}\frac{(m^3 - m)}{12}. \tag{2.11}$$

Thus the map $d_k \to L_k$ is a representation of the Virasoro algebra in B for $C = 1$. Moreover, this representation is unitary.

Proof of Proposition 2.3. From $\omega(a_n) = a_{-n}$ and the definition (2.9) it is easy to verify that $\omega(L_k) = L_{-k}$. Hence the Hermitian contravariant form defined in (2.6) is contravariant for *Vir*. Since the representation of $\mathscr{A}$ in B is unitary, the same holds for the representation of *Vir* with $C = 1$. We therefore only have to verify (2.11).

Lemma 2.2.

$$[a_k, L_n] = k a_{k+n} \quad (k, n \in \mathbb{Z}).$$ (2.12)

The following 'cutoff' procedure replaces calculations with infinite sums by calculations with finite sums.

Define the function ψ on $\mathbb{R}$ by:

$$\psi(x) = 1 \text{ if } |x| \leqslant 1; \quad \psi(x) = 0 \text{ if } |x| > 1.$$ (2.13a)

Put

$$L_n(\epsilon) = \frac{1}{2} \sum_{j \in \mathbb{Z}} {:} a_{-j} a_{j+n} {:} \psi(\epsilon j).$$ (2.13b)

Note that $L_n(\epsilon)$ contains only a finite number of terms if $\epsilon \neq 0$ and that $L_n(\epsilon) \to L_n$ as $\epsilon \to 0$. More precisely, the latter statement means that, given $v \in B$, $L_n(\epsilon)(v) = L_n(v)$ for ϵ sufficiently small.

Proof of Lemma 2.2. $L_n(\epsilon)$ as defined in (2.13b) differs from the same expression without normal ordering by a finite sum of scalars. This drops out of the commutator $[a_k, L_n(\epsilon)]$ and so

$$[a_k, L_n(\epsilon)] = \frac{1}{2} \sum_j [a_k, a_{-j} a_{j+n}] \psi(\epsilon j)$$

$$= \frac{1}{2} \sum_j [a_k, a_{-j}] a_{j+n} \psi(\epsilon j) + \frac{1}{2} \sum_j a_{-j} [a_k, a_{j+n}] \psi(\epsilon j)$$

$$= \frac{1}{2} k a_{k+n} \psi(\epsilon k) + \frac{1}{2} k a_{k+n} \psi(\epsilon(k+n))$$

using (2.2). The $\epsilon \to 0$ limit gives (2.12). ∎

End of proof of Proposition 2.3.

$$[L_m(\epsilon), L_n] = \frac{1}{2} \sum_j [a_{-j} a_{j+m}, L_n] \psi(\epsilon j)$$

$$= \frac{1}{2} \sum_j (-j) a_{n-j} a_{j+m} \psi(\epsilon j)$$

$$+ \frac{1}{2} \sum_j (j+m) a_{-j} a_{j+m+n} \psi(\epsilon j).$$

We split the first sum into terms satisfying $j \geqslant (n - m)/2$ which are in normal order and reverse the order of terms for which $j < (n - m)/2$ using the commutation relations. In the same way we split the second sum into terms satisfying $j \geqslant -(m + n)/2$ and $j < -(m + n)/2$. Then,

$$[L_m(\epsilon), L_n] = \frac{1}{2} \sum_j (-j) : a_{n-j} a_{j+m} : \psi(\epsilon j)$$

$$+ \frac{1}{2} \sum_j (j + m) : a_{-j} a_{j+m+n} : \psi(\epsilon j)$$

$$- \frac{1}{2} \delta_{m,-n} \sum_{j=-1}^{-m} j(m + j)\psi(\epsilon j) .$$

Making the transformation $j \to j + n$ in the first sum and taking the limit $\epsilon \to 0$, we get (2.11) since $-\frac{1}{2} \sum_{j=-1}^{-m} j(m + j) = (m^3 - m)/12$. ∎

We split the first sum into two sets satisfying $j > (n-m)/2$ which are normal ordered and reverse the order of terms for which $j < (n-m)/2$ using the commutation relations. In the same way we split the second sum into sets satisfying $j > (n+m)/2$ and $j < (n+m)/2$. Then

[illegible]

[illegible]

[illegible]

Making the transformation $j \to n - j$ in the first sum and taking the limit $\epsilon \to 0$, we get (2.11) since $\sum_{j=1}^{m} j(m+1-j) = (m^3-m)/12$. ∎

LECTURE 3

3.1. Complete reducibility of the oscillator representations of *Vir*

In the previous lecture we found unitary representations of the Virasoro algebra (2.1) with central charge $C = 1$ in terms of the operators L_k:

$$d_k \to L_k \ (k \in \mathbb{Z}), \quad C \to 1. \tag{3.1}$$

The L_k defined by (2.9) can be written without the normal ordering notation as

$$L_k = \frac{\epsilon}{2} a_{k/2}^2 + \sum_{j > -k/2} a_{-j} \, a_{j+k}, \tag{3.2}$$

where $\epsilon = 0$ if k is odd, $= 1$ if k is even.

The operators a_k form a unitary irreducible representation of the oscillator algebra $\mathscr{A}$ in the vector space B defined by (2.3). From (3.2) the 'energy operator' L_0 is given by

$$L_0 = \mu^2/2 + \sum_{j > 0} a_{-j} \, a_j. \tag{3.3}$$

From (2.7) we know that B can be written as a direct sum of finite dimensional subspaces of elements of degree j. It is easily seen that each such subspace is an eigenspace of L_0 with eigenvalue $\mu^2/2 + j$, where $j \in \mathbb{Z}_+$. Thus

$$B = \bigoplus_{j \in \mathbb{Z}_+} B_j, \tag{3.4}$$

19

where B_j is the $(\mu^2/2 + j)$-eigenspace of L_0 as well as the subspace of elements of degree j.

The decomposition (3.4) of the representation space as a direct sum of finite-dimensional subspaces is extremely useful as it enables us to use algebraic methods in an infinite-dimensional problem, just as in the previous lectures in the cases of $\eth$ and $\mathscr{A}$ which have similar decompositions.

The decomposition (3.4) can now be used to discuss whether this representation of Vir in B is irreducible. Lemma 1.1 applies to the present situation and we can conclude that any subrepresentation U of B will have the decomposition:

$$U = \bigoplus_{j \in \mathbb{Z}_+} (U \cap B_j). \tag{3.5}$$

We recall from Proposition 2.3 that the representation of Vir in B is also unitary and $\omega(L_0) = L_0$. It follows that the eigenspaces of L_0 which appear in (3.4) are mutually orthogonal with respect to the Hermitian contravariant form $\langle \cdot \mid \cdot \rangle$ on B. Given a subrepresentation U of B with the decomposition (3.5), then, denoting by U_j the finite-dimensional vector space $U \cap B_j$ we can define a subspace $U^\perp$ by

$$U^\perp = \bigoplus_{j \in \mathbb{Z}_+} U_j^\perp, \tag{3.6}$$

where $U_j^\perp$ is the finite-dimensional orthogonal complement of U_j in B_j. Clearly we have:

$$B = U \oplus U^\perp, \tag{3.7}$$

since

$$U^\perp = \{v \in B \mid \langle U \mid v \rangle = 0\}. \tag{3.8}$$

It is now clear that $U^\perp$ is also an invariant subspace for Vir, since $\langle U \mid U^\perp \rangle = 0$ and $L_j U \subset U$ implies $0 = \langle L_j U \mid U^\perp \rangle = \langle U \mid L_{-j} U^\perp \rangle$ and so $U^\perp$ is also invariant under the L_j $(j \in \mathbb{Z})$. We have proved:

Proposition 3.1. The unitary representation (3.1) of the Virasoro algebra with central charge $C = 1$ in the Fock space B is a direct sum of irreducible representations. ∎

3.2. Highest weight representations of *Vir*

We can construct a subrepresentation B' of B as follows. From the properties of the vacuum vector 1 listed in (2.4) and the definition (3.2) of the operators L_k, we see that

$$L_k(1) = 0 \quad (k > 0), \tag{3.9}$$

$$L_0(1) = h \cdot 1, \quad \text{where} \quad h = \mu^2/2. \tag{3.10}$$

Let B' be the linear span of vectors

$$L_{-i_k} \ldots L_{-i_2} L_{-i_1}(1), \tag{3.11}$$

where we take arbitrary finite sequences $0 < i_1 \leqslant i_2 \leqslant \ldots \leqslant i_k$. It is easy to see from (3.9), (3.10) and (3.11) that B' is invariant under the L_j and so we get a subrepresentation of B, called the *highest component* of B. Note that vectors (3.11) are not linearly independent in general.

Proposition 3.2. The representation in B' is an irreducible representation of *Vir*.

Proof. If B' is not irreducible, then, as above, it can be written as the direct sum of two representations: $B' = U \oplus U^{\perp}$. Each summand has a decomposition of the form (3.5). The vacuum vector 1 spans the h-eigenspace of L_0 and hence belongs to exactly one summand, which implies that all vectors of (3.11) are in this summand. Hence the other summand is 0 and B' is irreducible. ∎

Definition 3.1. A *highest weight representation* of *Vir* is a representation in a vector space V which admits a nonzero vector v such that for given complex numbers c, h:

$$C(v) = cv, \tag{3.12a}$$

$$d_0(v) = hv, \tag{3.12b}$$

and V is the linear span of vectors of the form

$$d_{-i_k} \ldots d_{-i_1}(v) \quad (0 < i_1 \leqslant \ldots \leqslant i_k). \tag{3.12c}$$

The pair (c, h) is called the *highest weight*, and v is called the *highest weight vector*.

Note that (3.12b, c) imply (see below):

$$d_i(v) = 0 \quad \text{for} \quad i > 0. \tag{3.12d}$$

Remark 3.1. B' is an example of a highest weight representation of *Vir* with highest weight vector 1 and highest weight $(1, \mu^2/2)$. Note that B' is spanned by elements of the form (3.12c) since $L_n(1) = 0$ for $n > 0$ and since B' is irreducible. ∎

Remark 3.2. It follows from the proof of Proposition 3.2 that a unitary highest weight representation of *Vir* is irreducible. ∎

Remark 3.3. While B' is certainly an irreducible representation of *Vir* for $c = 1$, the question remains as to whether or not $B' = B$. The answer depends on the value of μ. We shall see that $B' = B$ for generic values of μ, but this does not hold for special values. For example, if $\mu = 0$ then the vector $u = a_{-1}(1)$ satisfies (3.9) and (3.10) with $h = 1$. Hence we can use u to form a highest weight representation of *Vir* with $c = 1$ which is orthogonal to B'. ∎

Let V be a highest weight representation of *Vir* with highest weight (c, h). We observe that all vectors of the form (3.12c) with a fixed value of $j = i_1 + i_2 + \ldots + i_k$ span the eigenspace V_{h+j} of d_0 with eigenvalue $h + j$, so that we have:

$$V = \bigoplus_{j \in \mathbb{Z}_+} V_{h+j}. \tag{3.13}$$

It is clear that $\dim V_{h+j} \leqslant p(j)$ and that equality holds if and only if all such vectors are linearly independent. Note also that (3.12d) follows from (3.13).

The formal power series

$$\text{ch } V = \text{tr}_V \, q^{d_0} \equiv \sum_{j \in \mathbb{Z}_+} (\dim V_{h+j}) q^{h+j}$$

is called the *character* of the representation V of *Vir*.

3.3.   Verma representations $M(c, h)$ and irreducible highest weight representations $V(c, h)$ of *Vir*

Definition 3.2. If all the vectors of the form (3.12c) in a highest weight representation of *Vir* are linearly independent, then this highest weight representation is called a *Verma representation*.

We shall denote the Verma representation by $M(c, h)$, where c, h are defined by (3.12a, b).

Remark 3.4. We can conclude from Remark 3.3 that B' is a Verma representation for generic values of h, but not, in particular, for $h = 0$. ∎

Since the Verma representation $M(c, h)$ is the highest weight representation for given c, h in which all vectors of the form (3.12c) are linearly independent, it follows that there is a homomorphism from $M(c, h)$ to any other highest weight representation of *Vir* with the same values of c, h which maps highest weight vector to highest weight vector and commutes with the action of *Vir*. This implies that any highest weight representation is isomorphic to a quotient of the Verma representation. It follows, in particular, that (c, h) determines the Verma representation $M(c, h)$ uniquely.

The existence of a Verma representation can be established for any c, h by standard Lie algebra techniques. We start with the universal enveloping algebra U of *Vir*, form the left ideal $I(c, h)$ generated by the elements $\{d_n(n > 0), d_0 - h \cdot \mathbf{1}, C - c \cdot \mathbf{1}\}$ where $\mathbf{1}$ is the identity element of U, and let $M(c, h) = U/I(c, h)$. The algebra *Vir* acts on $M(c, h)$ by left multiplication so that $M(c, h)$ is a representation of *Vir*. Consider the identity element $\mathbf{1}$ of U and let v be its image in $M(c, h)$. Then for $n > 0$, $d_n(v) = 0$ and $C(v) = cv$, $d_0(v) = hv$. Thus $M(c, h)$ is a highest weight representation with highest weight (c, h). The linear independence of vectors of the form (3.12c) is assured by the Poincaré-Birkhoff-Witt theorem. Hence $M(c, h)$ is a Verma representation.

Proposition 3.3. (a) The Verma representation $M(c, h)$ has the decomposition

$$M(c, h) = \bigoplus_{k \in \mathbb{Z}_+} M(c, h)_{h+k} \tag{3.14}$$

where $M(c, h)_{h+k}$ is the $(h + k)$-eigenspace of d_0 of dimension $p(k)$ spanned by vectors of the form

$$d_{-i_s} \ldots d_{-i_1}(v) \quad \text{with} \quad 0 < i_1 \leqslant \ldots \leqslant i_s, i_1 + \ldots + i_s = k.$$

One has:

$$\operatorname{ch} M(c, h) = q^h / \varphi(q), \tag{3.15}$$

where $\varphi(q)$ is defined by (2.8b).

(b) $M(c, h)$ is indecomposable, i.e. we cannot find nontrivial subrepresentations V, W such that

$$M(c, h) = V \oplus W. \tag{3.16}$$

(c) $M(c, h)$ has a unique maximal proper subrepresentation $J(c, h)$, and

$$V(c, h) = M(c, h)/J(c, h)$$

is the unique irreducible highest weight representation with highest weight (c, h). We have:

$$\operatorname{ch} V(c, h) \leqslant q^h / \varphi(q). \tag{3.17}$$

Proof.   (a) follows from the previous discussion.

(b) Lemma 1.1 applies so that, if (3.16) were true, both V and W would be graded according to (3.14) and therefore the vacuum vector v would belong to V or W. If v belongs to a subrepresentation, then this subrepresentation must coincide with $M(c, h)$.

(c) By Lemma 1.1 all proper subrepresentations are graded according to (3.14) and so is their sum which is also a proper subrepresentation since it does not contain v. The maximal subrepresentation $J(c, h)$ is therefore the sum of all proper subrepresentations. The rest follows immediately. ∎

Note that the irreducible highest weight representation $V(c, h)$ of *Vir* can be defined as the unique irreducible representation having a vector v such that

$$d_i(v) = 0 \quad \text{for} \quad i > 0, \quad d_0(v) = hv, \quad c(v) = cv. \tag{3.18}$$

Similarly, the Verma representation $M(c, h)$ can be defined as the unique representation of *Vir* having a vector v satisfying (3.18) and such that all monomials (3.12c) are linearly independent.

Remark 3.5. Note that the representations $V(c, h)$ with $h \geqslant 0$ are precisely all irreducible positive-energy representations of *Vir*. ∎

The antilinear anti-involution ω of Vir extends to an antilinear anti-involution of its universal enveloping algebra U. From Proposition 3.3(a) it makes sense to define for each $u \in M(c,h)$ its *expectation value* $\langle u \rangle$ as the coefficient of the highest weight vector v in the expansion of u with respect to (3.14) (since $M(c,h)_h = \mathbb{C}v$). Since any element of U is a linear combination of elements of the form

$$R = (d_{-j_1} \ldots d_{-j_s})(d_0^k C^r)(d_{i_1} \ldots d_{i_t}),$$

where $s, t \geqslant 0$, $j_1 \geqslant \ldots \geqslant j_s > 0$, $i_1 \geqslant \ldots \geqslant i_t > 0$, and since

$$\omega(R) = (d_{-i_t} \ldots d_{-i_1})(d_0^k C^r)(d_{j_s} \ldots d_{j_1}),$$

we deduce the following important property of the expectation value:

$$\langle \omega(R)(v) \rangle = \overline{\langle R(v) \rangle} \quad \text{provided that} \quad c, h \in \mathbb{R}. \tag{3.19}$$

Proposition 3.4. (a) Provided that c and h are real, $M(c,h)$ carries a unique contravariant Hermitian form $\langle \cdot | \cdot \rangle$ such that $\langle v | v \rangle = 1$, where v is the highest weight vector.
(b) The eigenspaces of d_0 are pairwise orthogonal.
(c) $\mathrm{Ker}\langle \cdot | \cdot \rangle = J(c,h)$. Hence $V(c,h)$ carries a unique contravariant Hermitian form such that $\langle v | v \rangle = 1$ and this form is nondegenerate.

Proof. (a), (b) We define $\langle \cdot | \cdot \rangle$ on monomials $P(v) = d_{-i_m} \ldots d_{-i_1}(v)$ and $Q(v) = d_{-j_n} \ldots d_{-j_1}(v)$ by:

$$\langle P(v) | Q(v) \rangle = \langle \omega(P)Q(v) \rangle = \langle d_{i_1} \ldots d_{i_m} d_{-j_n} \ldots d_{-j_1}(v) \rangle. \tag{3.20}$$

This is a Hermitian form due to (3.19). This Hermitian form is obviously contravariant. Clearly (3.20) vanishes if $i_1 + \ldots + i_m \neq j_1 + \ldots + j_n$.
(c) $\mathrm{Ker}\langle \cdot | \cdot \rangle \equiv \{u \in M(c,h) \,|\, \langle u | w \rangle = 0 \text{ for all } w \in M(c,h)\}$ clearly is a proper subrepresentation of $M(c,h)$ since the highest weight vector v is not contained in it. Moreover any proper subrepresentation W of $M(c,h)$ lies in $\mathrm{Ker}\langle \cdot | \cdot \rangle$. This is clear from (3.20), since if $P(v) \in W$ and $Q(v) \in M(c,h)$, then $\omega(Q)P(v) \in W$, and if $\langle \omega(Q)P(v) \rangle \neq 0$ we get, using Lemma 1.1, that $v \in W$ and hence $W = M(c,h)$. $\blacksquare$

Proposition 3.5. There exists at most one unitary highest weight representation on Vir for a given highest weight (c,h), *viz.* $V(c,h)$. A necessary condition for the unitarity of $V(c,h)$ is that $c \geqslant 0$, $h \geqslant 0$.

Proof. The first statement follows immediately from Remark 3.2 and Proposition 3.4. A necessary condition for unitarity is that

$$c_n = \langle d_{-n}v \,|\, d_{-n}v \rangle \geqslant 0 \quad \text{for each } n \geqslant 0 .$$

But contravariance and the commutation rules show that

$$c_n = 2nh + c(n^3 - n)/12 . \tag{3.21}$$

Putting $n = 1$, we get $c_1 = 2h$ so that we must have $h \geqslant 0$. Moreover, (3.21) shows that c_n is dominated by cn^3 for large n, so that $c \geqslant 0$ is also necessary. ∎

Remark 3.6. The Hilbert completion of every unitary representation $V(c, h)$ can be integrated to a projective representation of Diff S^1 (see Goodman-Wallach [1985]). ∎

3.4. More (unitary) oscillator representations of *Vir*

We have constructed a unitary representations of *Vir* for $c = 1, h \geqslant 0$ using the oscillator representation. Now if $\mathfrak{g}$ is a Lie algebra with representations in V_1 and V_2, then it has a representation in $V_1 \otimes V_2$ defined by

$$a(v_1 \otimes v_2) = (av_1) \otimes v_2 + v_1 \otimes (av_2) \tag{3.22}$$

for every a in $\mathfrak{g}$. Moreover, if V_1 and V_2 are unitary, we can define a Hermitian form on $V_1 \otimes V_2$ by

$$\langle v_1 \otimes v_2 \,|\, w_1 \otimes w_2 \rangle = \langle v_1 \,|\, w_1 \rangle \langle v_2 \,|\, w_2 \rangle \tag{3.23}$$

and it is easy to verify that this is a positive-definite Hermitian contravariant form. Hence the tensor product of two unitary representations of $\mathfrak{g}$ is unitary. Taking tensor products of the oscillator representation with itself we can construct oscillator representations of *Vir* for any positive integral value of c and any $h \geqslant 0$. Taking the highest component, we have a unitary highest weight representation of *Vir* for any $c = 1, 2, \dots$ and $h \geqslant 0$.

The following modification of the Virasoro construction was found by Fairlie (see Chodos-Thorn [1974]). Defining for arbitrary real λ, μ

$$\tilde{L}_0 = (\mu^2 + \lambda^2)/2 + \sum_{j>0} a_{-j}a_j \tag{3.24a}$$

and, for $k \neq 0$

$$\tilde{L}_k = \frac{1}{2} \sum_{j \in \mathbb{Z}} a_{-j} a_{j+k} + i\lambda k a_k , \qquad (3.24b)$$

we can easily verify, using (2.11) and (2.12), that

$$[\tilde{L}_m, \tilde{L}_n] = (m - n)\tilde{L}_{m+n} + \delta_{m,-n} \frac{(m^3 - m)}{12}(1 + 12\lambda^2) . \qquad (3.25)$$

(As before, we take $\mathbb{1} = 1$ and $a_0 = \mu$.) Clearly $\omega(\tilde{L}_k) = \tilde{L}_{-k}$ and the form $\langle \cdot | \cdot \rangle$ defined on the Fock space B is a Hermitian contravariant form for this representation of Vir as well. Taking the highest component, we therefore have a unitary highest weight representation of Vir for $c = 1 + 12\lambda^2$, $h = (\lambda^2 + \mu^2)/2$. In the $c \geqslant 0, h \geqslant 0$ quadrant we therefore have unitarity of $V(c, h)$ for points (c, h) lying in the region between the line $c = 1$ and $h = (c - 1)/24$. Taking into account the possibility of tensoring, the situation is as summarized in Figure 3.1. The entire region to the right of the line $c = 1$ is a region of unitarity, with the exception of the shaded triangles. We shall see later that $V(c, h)$ is unitary here as well, but we do not have a manifestly unitary oscillator construction. (This is a very interesting open problem.)

The following notion is important in the analysis of irreducibility.

Definition 3.3. A vector u of a representation of Vir in V is called *singular* if it is nonzero and

$$d_n(u) = 0 \quad \text{for} \quad n > 0 . \qquad (3.26)$$

Of course a scalar multiple of the highest weight vector of a highest weight representation is singular. The following assertion is clear by Lemma 1.1 and Proposition 3.3.

Proposition 3.6. A highest weight representation of Vir is irreducible if and only if it has no singular vectors other than scalar multiples of the highest weight vector. ∎

We have not said anything about the region $0 < c < 1$, $h \geqslant 0$. We shall now show that there is an oscillator representation for the Virasoro algebra for $c = 1/2$. The oscillators will, however, be fermionic oscillators

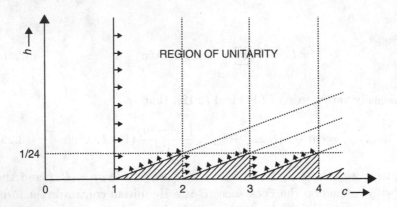

Figure 3.1

ψ_n satisfying the anticommutation relations

$$[\psi_m, \psi_n]_+ \equiv \psi_m \psi_n + \psi_n \psi_m = \delta_{m,-n} \quad (m, n \in \delta + \mathbb{Z}), \qquad (3.27a)$$

where either $\delta = 0$ ('Ramond sector') or $\delta = 1/2$ ('Neveu-Schwarz sector'). In the Ramond sector ($\delta = 0$), $m = n = 0$ is allowed and we have

$$\psi_0^2 = \frac{1}{2}. \qquad (3.27b)$$

The algebra (3.27) can be represented in the vector space

$$V_\delta = \bigwedge[\xi_i \,|\, i \geqslant 0, \ i \in \delta + \mathbb{Z}], \qquad (3.28)$$

where the symbol $\bigwedge$ on the right means the exterior algebra generated by the ξ_i. Thus V_δ is the direct sum of $\mathbb{C}$ and antisymmetric tensors in the ξ_i of rank n for every $n \geqslant 1$. The antisymmetry is taken care of by the assumed relations

$$\xi_i \xi_j = -\xi_j \xi_i \qquad (3.29)$$

for all i, j. The algebra (3.27) is represented in V_δ by the identifications

$$\left.\begin{array}{c} \psi_n \to \partial/\partial\xi_n \\[4pt] \psi_{-n} \to \xi_n \end{array}\right\} \quad (n > 0). \qquad (3.30)$$

For $n = 0$, which occurs in the Ramond sector only,

$$\psi_0 \to \frac{1}{\sqrt{2}}(\xi_0 + \partial/\partial\xi_0). \qquad (3.31)$$

We shall identify the ψ_n with the corresponding operators in V_δ. The antilinear anti-involution ω, the vacuum expectation value $\langle \; \rangle$ and the Hermitian contravariant form $\langle \cdot | \cdot \rangle$ can be defined exactly as for the oscillator algebra. In fact, the monomials $\xi_{i_1} \ldots \xi_{i_s}$ $(i_1 < i_2 < \ldots < i_s)$ form an orthonormal basis for $\langle \cdot | \cdot \rangle$, and therefore this representation is unitary.

Proposition 3.7. Let L_k $(k \in \mathbb{Z})$ be the operators in V_δ defined by

$$L_k = \delta_{k,0}\frac{(1-2\delta)}{16} + \frac{1}{2}\sum_j j : \psi_{-j}\psi_{j+k} : \qquad (3.32)$$

where j runs over $\delta + \mathbb{Z}$ and the normal ordering is defined by

$$\begin{aligned} :\psi_j\psi_k: &= \psi_j\psi_k \text{ if } k \geqslant j \\ &= -\psi_k\psi_j \text{ if } k < j. \end{aligned} \qquad (3.33)$$

Then:

(i) $[\psi_m, L_k] = (m + k/2)\psi_{m+k}$, $\qquad (3.34)$

(ii) $[L_m, L_n] = (m-n)L_{m+n} + \delta_{m,-n}\dfrac{(m^3 - m)}{24}$. $\qquad (3.35)$

Proof. A straightforward calculation using the same method as for the oscillator algebra in Lecture 2. ∎

From (3.32) we see that

$$L_0 = (1 - 2\delta)/16 + \sum_{j>0} j\,\psi_{-j}\psi_j \quad (j \in \delta + \mathbb{Z}). \qquad (3.36)$$

The representation of Vir in V_δ has central charge $c = 1/2$. It is not irreducible: the subspaces V_δ^+ and V_δ^- of even and odd elements respectively are subrepresentations of V_δ. The elements of lowest energy in $V_{1/2}^+$ and $V_{1/2}^-$ are 1 and $\xi_{1/2}$ with energy 0 and 1/2 respectively, and in V_0^+ and V_0^- are 1 and ξ_0, both of energy 1/16 (as is easily seen from (3.36)).

Furthermore, all four representations $V_\delta^\pm$ are irreducible. Indeed, by Proposition 3.6, if it were not the case, $V_\delta^\pm$ would contain a singular vector u of energy $h_0 + n_0$ where $h_0 = 0, 1/2$ or $1/16$ and $n_0 \geqslant 1$. The vector u would generate a unitary representation of Vir with highest weight $(1/2, h + n_0)$.

But, as we shall see in Lecture 12, the representations $V(1/2, h)$ are unitary for $h = 0$, $1/16$ and $1/2$ only. Thus, we arrive at the following

Proposition 3.8. The representation of Vir given in Proposition 3.7 is irreducible in the even and odd subspaces $V_\delta^\pm$. In the Neveu-Schwarz sector we have a unitary highest weight representation of Vir with $c = 1/2$, $h = 0$ in $V_{1/2}^+$ and with $c = 1/2$, $h = 1/2$ in $V_{1/2}^-$. In the Ramond sector we have a unitary highest weight representation of Vir with $c = 1/2$, $h = 1/16$ in both V_0^+ and V_0^-. Finally we have:

$$\operatorname{ch} V\left(\frac{1}{2}, 0\right) \pm \operatorname{ch} V\left(\frac{1}{2}, \frac{1}{2}\right) = \prod_{n \in \mathbb{Z}_+} (1 \pm q^{n+1/2}), \qquad (3.37)$$

$$\operatorname{ch} V\left(\frac{1}{2}, \frac{1}{16}\right) = q^{1/16} \prod_{n \in \mathbb{Z}_+} (1 + q^{n+1}). \qquad (3.38)$$

       ∎

So far we have constructed two types of irreducible representations of Vir: the representations $V'_{\alpha,\beta}$ and the highest weight representations $V(c, h)$. Of course, we also have the lowest weight representations $V^*(c, h)$, dual to $V(c, h)$.

Conjecture 3.1. (Kac [1982]). An irreducible representation of Vir for which the energy operator is diagonalizable with finite-dimensional eigenspaces is either $V'_{\alpha,\beta}$ or $V(c, h)$ or $V^*(c, h)$.

This conjecture has been checked by Kaplansky and Santharoubane [1985] in the case when all eigenspaces of d_0 have dimension $\leqslant 1$. Chari and Pressley [1987] proved that the conjecture is true for unitary representations. The conjecture was open at the time of the first edition. It was proved later in full generality by Mathieu [1992] using reduction mod p methods.   ∎

LECTURE 4

4.1. Lie algebras of infinite matrices

In this lecture we shall study Lie algebras of infinite matrices and realize the algebras discussed earlier as its subalgebras. We shall then see how Dirac's positron theory can be given a representation-theoretic interpretation and used to obtain highest weight (= positive-energy) representations of these Lie algebras.

Let

$$V = \bigoplus_{j \in \mathbb{Z}} \mathbb{C}v_j \tag{4.1}$$

be an infinite-dimensional complex vector space with a fixed basis $\{v_j \,|\, j \in \mathbb{Z}\}$. We shall identify v_j with the column vector with 1 as the j-th entry and 0 elsewhere. Any vector in V has only a finite, but arbitrary, number of nonzero coordinates $(a_i)_{i \in \mathbb{Z}}$; this identifies V with $\mathbb{C}^\infty$, the space of such column vectors.

The Lie algebra $g\ell_\infty$ is defined by:

$$g\ell_\infty = \{(a_{ij})_{i,j \in \mathbb{Z}} \,|\, \text{all but a finite number of the } a_{ij} \text{ are } 0\}, \tag{4.2}$$

with the Lie bracket being the ordinary matrix commutator.

We denote by E_{ij} the matrix with 1 as the (i,j) entry and all other entries 0. The E_{ij} $(i,j \in \mathbb{Z})$ form a basis for $g\ell_\infty$. Clearly,

$$E_{ij}v_k = \delta_{jk}v_i \tag{4.3a}$$

and

$$E_{ij}E_{mn} = \delta_{jm}E_{in}. \tag{4.3b}$$

The commutation relations of $g\ell_\infty$ can be expressed as the commutation relations of the E_{ij}:

$$[E_{ij}, E_{mn}] = \delta_{jm} E_{in} - \delta_{ni} E_{mj}. \tag{4.4}$$

The Lie algebra $g\ell_\infty$ can be viewed as the Lie algebra of the group GL_∞ defined as follows:

$$GL_\infty = \{A = (a_{ij})_{i,j \in \mathbb{Z}} \mid A \text{ invertible and all but a}$$
$$\text{finite number of } a_{ij} - \delta_{ij} \text{ are } 0\}. \tag{4.5}$$

The group operation is matrix multiplication.

We define a bigger Lie algebra $\bar{\mathfrak{a}}_\infty$:

$$\bar{\mathfrak{a}}_\infty = \{(a_{ij}) \mid i, j \in \mathbb{Z}, \, a_{ij} = 0 \text{ for } |i - j| \gg 0\}. \tag{4.6}$$

Matrices in $\bar{\mathfrak{a}}_\infty$ have a finite number of nonzero diagonals. It is easy to see that the product of two matrices in $\bar{\mathfrak{a}}_\infty$ is well defined, and is again in $\bar{\mathfrak{a}}_\infty$, so that $\bar{\mathfrak{a}}_\infty$ is a Lie algebra with the matrix commutator, containing $g\ell_\infty$ as a subalgebra.

We define the shift operators Λ_k by

$$\Lambda_k v_j = v_{j-k}. \tag{4.7}$$

Clearly, by (4.3a),

$$\Lambda_k = \sum_{i \in \mathbb{Z}} E_{i, i+k}. \tag{4.8}$$

Λ_k is the matrix with 1 at each entry on the k-th diagonal ($k = 0$ being the principal diagonal) and 0 elsewhere. The Λ_k form a commutative subalgebra of $\bar{\mathfrak{a}}_\infty$:

$$[\Lambda_j, \Lambda_k] = 0 \quad (j, k \in \mathbb{Z}). \tag{4.9}$$

In Lecture 1 we found representations of $\mathfrak{d}$ in $V_{\alpha,\beta}$ given by equation (1.9). We shall change our notation by replacing v_k in (1.9) by v_{-k} so that (1.9) becomes:

$$d_n(v_k) = (k - \alpha - \beta(n+1))v_{k-n}, \tag{4.10}$$

from which we deduce that

$$d_n = \sum_{k \in \mathbb{Z}} (k - \alpha - \beta(n+1)) E_{k-n,k}. \tag{4.11}$$

Clearly the d_n are in $\bar{\mathfrak{a}}_\infty$ since d_n has nonzero entries only on the n-th diagonal. This gives an inclusion of $\mathfrak{d}$ as a subalgebra of $\bar{\mathfrak{a}}_\infty$.

4.2. Infinite wedge space F and the Dirac positron theory

In trying to construct a quantum theory of a single electron with positive energy, Dirac was led to construct a multiparticle theory of electrons and positrons. Let us recall the description given in his book 'The Theory of Quantum Mechanics' (Dirac [1958]):

"... the wave equation for the electron admits of twice as many solutions as it ought to, half of them referring to states with negative values for the kinetic energy ... we are led to infer that the negative-energy solutions ... refer to the motion of a new kind of particle having the mass of an electron and the opposite charge. Such particles have been observed experimentally and are called positrons. ... We assume that nearly all the negative-energy states are occupied, with one electron in each state in accordance with the exclusion principle of Pauli. An unoccupied negative-energy state will now appear as something with a positive energy, since to make it disappear, i.e. to fill it up, we should have to add to it an electron with negative energy. We assume that these unoccupied negative-energy states are the positrons.

These assumptions require there to be a distribution of electrons of infinite density everywhere in the world. A perfect vacuum is a region where all the states of positive energy are unoccupied and all those of negative energy are occupied ... the infinite distribution of negative-energy electrons does not contribute to the electric field ... there will be a contribution $-e$ for each occupied state of positive energy and a contribution $+e$ for each unoccupied state of negative energy.

The exclusion principle will operate to prevent a positive-energy electron ordinarily from making transitions to states of negative energy. It will still be possible, however, for such an electron to drop into an unoccupied state of negative energy. In this case we should have an electron and positron disappearing simultaneously, their energy being emitted in the form of radiation. The converse process would consist in the creation of an electron and a positron from electromagnetic radiation."

An electron is described by its vector space of states, which we shall take to be the vector space V defined in (4.1). We shall call v_j the state of an electron of energy $j \in \mathbb{Z}$. The energy is thus not always positive. To fix this, Dirac theory requires us to consider an infinite number of such electrons satisfying the Pauli exclusion principle. We must therefore consider the *infinite wedge space* $\bigwedge^\infty V$, where the symbol $\bigwedge$ stands for the exterior product, i.e., the antisymmetric tensor product. We can now define the vacuum state in accordance with Dirac theory as the state with positive energy states empty, but all negative energy states occupied. Denoting by $\bigwedge$ the exterior product of vectors, we have the perfect vacuum:

$$\psi_0 = v_0 \wedge v_{-1} \wedge v_{-2} \wedge \dots . \tag{4.12}$$

Note that we have included the zero-energy state with the negative energy states. All states are now produced by finite excitations of the perfect vacuum which produce simultaneously an electron with positive energy and a hole, i.e. an unoccupied state, in the negative-energy "sea". We define the state space

$$F^{(0)} = \bigwedge_{(0)}^\infty V \tag{4.13}$$

as the vector space with basis consisting of elements of the form

$$\psi = v_{i_0} \wedge v_{i_{-1}} \wedge v_{i_{-2}} \wedge \dots , \tag{4.14}$$

where

(i) $i_0 > i_{-1} > \dots$ $\qquad\qquad$ (4.15a)

(ii) $i_k = k$ for $k \ll 0$. $\qquad\qquad$ (4.15b)

Conditions (4.15) ensure that ψ has an equal number of electrons and holes (positrons) so that ψ is a (charge conserving) excitation of the vacuum state ψ_0. We can compute the degree of the excitation of ψ when its labels satisfy (4.15a) by subtracting from each label of ψ the corresponding label of ψ_0. This leads us to define the *degree* of ψ (or the *energy*) as

$$\deg \psi = \sum_{s=0}^{\infty} (i_{-s} + s). \tag{4.16}$$

The degree of each ψ in $F^{(0)}$ is a finite nonnegative integer because of (4.15b). Now let k be an arbitrary positive integer and let $\{k_0, k_1, \dots, k_{n-1}\}$ be a partition of k in non-increasing order, i.e.

$$k = k_0 + k_1 + \dots + k_{n-1} \tag{4.17a}$$

where

$$k_0 \geqslant k_1 \geqslant \ldots \geqslant k_{n-1}. \qquad (4.17\text{b})$$

Then this partition of k defines a unique ψ satisfying (4.15), *viz.*

$$\psi = v_{j_0} \wedge v_{j_{-1}} \wedge \ldots \wedge v_{j_{-n+1}} \wedge v_{-n} \wedge v_{-n-1} \wedge \ldots \qquad (4.18\text{a})$$

where

$$j_{-i} = k_i - i \quad (i = 0, \ldots, n-1). \qquad (4.18\text{b})$$

This leads immediately to the following proposition:

Proposition 4.1. Let $F_k^{(0)}$ denote the linear span of all vectors of degree k. Then

(a) $F^{(0)} = \bigoplus_{k \in \mathbb{Z}_+} F_k^{(0)}, \quad F_0^{(0)} = \mathbb{C}\psi_0$ \hfill (4.19)

(b) $\dim F_k^{(0)} = p(k)$ \hfill (4.20)

(c) $\dim_q F^{(0)} \equiv \sum_k (\dim F_k^{(0)}) q^k = 1/\varphi(q),$ \hfill (4.21)

where $\varphi(q)$ is defined by (2.8b). ∎

Remark 4.1. An alternative way of computing $\deg \psi$ is by the Dirac recipe:

$$\deg \psi = \sum_s (i_s > 0 \text{ which occur}) - \sum_s (i_s \leqslant 0 \text{ which do not occur}). \quad (4.22)$$

∎

The space $F^{(0)}$ is constructed starting from a particular reference vector, the perfect vacuum ψ_0. We can consider a larger vector space $F = \bigwedge^\infty V$ with the basis consisting of all elements of the form (4.14) with labels satisfying (4.15a), with the sole restriction of Dirac theory that there should be only a finite number of unoccupied negative energy states (holes). Such elements will be called *semi-infinite monomials*. F can be decomposed as the vector sum of subspaces $F^{(m)}$:

$$F = \bigoplus_{m \in \mathbb{Z}} F^{(m)} \qquad (4.23)$$

as follows. Each $F^{(m)}$ is based on a reference vector (m-th vacuum)

$$\psi_m = v_m \wedge v_{m-1} \wedge v_{m-2} \wedge \ldots . \tag{4.24}$$

The space $F^{(m)}$ is the linear span of semi-infinite monomials of the form

$$\psi = v_{i_m} \wedge v_{i_{m-1}} \wedge \ldots , \tag{4.25}$$

where

(i) $i_m > i_{m-1} > \ldots$ $\hspace{3cm}$ (4.26a)

(ii) $i_k = k$ for $k \ll 0$. $\hspace{3cm}$ (4.26b)

We can define

$$\deg \psi = \sum_{s=0}^{\infty} (i_{m-s} + s - m) \tag{4.27}$$

$$= \sum_s (i_s > m \text{ which occur}) - \sum_s (i_s \leqslant m \text{ which do not occur}). \tag{4.28}$$

Physicists call m the *charge number*.

Corollary 4.1. Each $F^{(m)}$ ($m \in \mathbb{Z}$) has a decomposition into subspaces $F_k^{(m)}$ of fixed degree $k \in \mathbb{Z}_+$ as in (4.19). The dimensions of these subspaces satisfy (4.20) and (4.21). ∎

4.3.   Representations of GL_∞ and $g\ell_\infty$ in F. Unitarity of highest weight representations of $g\ell_\infty$

We can define representations R of GL_∞ and r of $g\ell_\infty$ in F by:

$$R(A)(v_{i_1} \wedge v_{i_2} \wedge \ldots) = Av_{i_1} \wedge Av_{i_2} \wedge \ldots , \tag{4.29}$$

$$r(a)(v_{i_1} \wedge v_{i_2} \wedge \ldots) = av_{i_1} \wedge v_{i_2} \wedge \ldots$$
$$+ v_{i_1} \wedge av_{i_2} \wedge \ldots + \ldots . \tag{4.30}$$

Equations (4.29) and (4.30) are related by:

$$\exp(r(a)) = R(\exp a) , \quad a \in g\ell_\infty . \tag{4.31}$$

By (4.30) the basis elements of $g\ell_\infty$, viz. the E_{ij} are represented by $r(E_{ij})$, where

$$r(E_{ij})v_{i_1} \wedge v_{i_2} \wedge \ldots = 0 \quad \text{if} \quad j \notin \{i_1, i_2, \ldots\},$$

and $\qquad = v_{i_1} \wedge \ldots \wedge v_{i_{k-1}} \wedge v_i \wedge v_{i_{k+1}} \wedge \ldots$ if $j = i_k$.

$$(4.32)$$

Of course the right-hand side of (4.32) vanishes if the label i is repeated. The $r(E_{ij})$ obey (4.4) and map each $F^{(m)}$ into itself so that the representation r of $g\ell_\infty$ in F is a direct sum of representations r_m in each $F^{(m)}$. Recall that $F^{(m)}$ is the linear span of semi-infinite monomials of the form

$$\psi = v_{i_m} \wedge \ldots \wedge v_{i_{m-k}} \wedge v_{m-k-1} \wedge \ldots , \qquad (4.33a)$$

which, using (4.32), is

$$\psi = r(E_{i_m,m}) \ldots r(E_{i_{m-k},m-k})\psi_m . \qquad (4.33b)$$

Define a positive definite Hermitian form $\langle \cdot \, | \, \cdot \rangle$ on F by declaring semi-infinite monomials to be an orthonormal basis. Let ω be the standard antilinear anti-involution of $g\ell_\infty$:

$$\omega(a) = a^\dagger , \qquad (4.34)$$

where $a^\dagger$ denotes the transpose complex conjugate of the matrix a. One checks immediately (on the E_{ij}) that

$$\langle r(a)\psi \, | \, \psi' \rangle = \langle \psi \, | \, r(a^\dagger)\psi' \rangle , \qquad (4.35)$$

so that the form $\langle \cdot \, | \, \cdot \rangle$ is contravariant and the representation r of $g\ell_\infty$ on F is unitary. The decomposition (4.23) is clearly orthogonal. Moreover, it follows now from (4.33b) (by the argument proving Proposition 3.2) that all representations r_m are irreducible. Thus, we have proved

Proposition 4.2. The representation r of $g\ell_\infty$ in F is a direct sum of irreducible unitary representations r_m in $F^{(m)}$. ∎

The representation r was constructed by Kac and Peterson [1981] in a more general framework and was called the *infinite wedge representation*.

Each $F^{(m)}$ has a vector space decomposition into subspaces of fixed degree as noted in Corollary 4.1:

$$F^{(m)} = \bigoplus_{k \geq 0} F_k^{(m)} . \qquad (4.36)$$

We can determine the action of $g\ell_\infty$ on this decomposition by examining the action of its basis element E_{ij} given in (4.32), where we see that E_{ij} either replaces v_j by v_i or gives zero. The replacement of v_j by v_i changes the degree of the vector by $i - j$. Thus

$$r(E_{ij})F_k^{(m)} \subset F_{k+i-j}^{(m)}. \qquad (4.37)$$

If we define

$$\deg E_{ij} = i - j\,,$$

we can decompose $g\ell_\infty$ as the vector sum of homogeneous components $\mathfrak{g}_j$ of degree j:

$$g\ell_\infty = \underset{j\in\mathbb{Z}}{\oplus}\, \mathfrak{g}_j\,. \qquad (4.38)$$

A matrix in $\mathfrak{g}_j$ has nonzero entries only on the $|\,j\,|$-th diagonal above $(j < 0)$ or below $(j > 0)$ the principal diagonal. By (4.37):

$$r(\mathfrak{g}_j)F_k^{(m)} \subset F_{k+j}^{(m)}\,. \qquad (4.39)$$

Moreover,

$$r(\mathfrak{g}_j)\psi_m = 0 \ \ \text{for} \ \ j < 0, \qquad (4.40)$$

since

$$r(E_{ij})\psi_m = 0 \ \ \text{for} \ \ i < j\,.$$

Using (4.33b), we have:

$$F_k^{(m)} = \sum_{\substack{j_1+\ldots+j_n=k \\ j_1,\ldots,j_n\in\mathbb{Z}_+}} r_m(\mathfrak{g}_{j_1})\ldots r_m(\mathfrak{g}_{j_n})\psi_m\,. \qquad (4.41)$$

The vector space decomposition (4.38) is called the *principal gradation* of $g\ell_\infty$. The alternative definition (4.41) of the gradation of $F^{(m)}$ gives us a representation-theoretic interpretation of Dirac's definition of energy.

Let $\mathfrak{n}_+$ be the subalgebra of $g\ell_\infty$ consisting of strictly upper triangular matrices. Clearly,

$$\mathfrak{n}_+ = \underset{j<0}{\oplus}\, \mathfrak{g}_j\,. \qquad (4.42)$$

Then from (4.40) and (4.42):

$$r_m(\mathfrak{n}_+)\psi_m = 0\,, \tag{4.43a}$$

$$r_m(E_{ii})\psi_m = \lambda_i\psi_m\,, \tag{4.43b}$$

where

$$\begin{aligned}\lambda_i &= 1 \ \text{ if } \ i \leqslant m \\ &= 0 \ \text{ if } \ i > m\,.\end{aligned} \tag{4.44}$$

Definition 4.1. Given a collection of numbers $\lambda = \{\lambda_i \mid i \in \mathbb{Z}\}$, called a *highest weight*, we define the *irreducible highest weight representation* π_λ of the Lie algebra $g\ell_\infty$ as an irreducible representation on a vector space $L(\lambda)$ which admits a nonzero vector v_λ, called a highest weight vector, such that

$$\pi_\lambda(\mathfrak{n}_+)v_\lambda = 0\,, \tag{4.45}$$

$$\pi_\lambda(E_{ii})v_\lambda = \lambda_i v_\lambda\,. \tag{4.46}$$

Note the analogy of this definition with the definition (3.18) of an irreducible highest weight representation of *Vir*. In particular, the same argument as in Lecture 3 shows that $L(\lambda)$ is determined by λ.

Thus for each $m \in \mathbb{Z}$ we have constructed an irreducible highest weight representation r_m of $g\ell_\infty$ with highest weight

$$\omega_m = \{\lambda_i = 1 \ \text{ for } \ i \leqslant m\,, \ \lambda_i = 0 \ \text{ for } \ i > m\}\,. \tag{4.47}$$

The r_m are called the *fundamental representations* of $g\ell_\infty$ and the ω_m the *fundamental weights*. Thus F is a direct sum of all fundamental representations of $g\ell_\infty$.

We showed in Lecture 3 (see equations (3.22), (3.23)) that the tensor product of two unitary representations V_1 and V_2 is also unitary. Furthermore, if V_1 and V_2 are irreducible unitary highest weight representations with highest weight vectors v_1 and v_2, then the vector $v_1 \otimes v_2$ is a highest weight vector of an irreducible subrepresentation of $V_1 \otimes V_2$, its *highest component*. It has highest weight equal to the sum of the two highest weights. Thus, we have proved the following proposition.

Proposition 4.3. The irreducible highest weight representations of $g\ell_\infty$ with highest weight of the form $\sum_i k_i \omega_i$, where the k_i are nonnegative integers, are unitary. ∎

It is easy to see that the unitarity of a highest weight representation of $g\ell_\infty$ with highest weight $\sum k_i \omega_i$ forces the k_i to be nonnegative integers (cf. Lecture 9).

Before concluding this subsection, let us note the following formula for the representation R_m of $A \in GL_\infty$ on $F^{(m)}$:

$$R_m(A)(v_{i_m} \wedge v_{i_{m-1}} \wedge \ldots) = \sum_{j_m > j_{m-1} > \ldots} \left(\det A^{i_m, i_{m-1}, \ldots}_{j_m, j_{m-1}, \ldots} \right)$$

$$\times v_{j_m} \wedge v_{j_{m-1}} \wedge v_{j_{m-2}} \wedge \ldots , \qquad (4.48)$$

where $A^{i_m, i_{m-1}, \ldots}_{j_m, j_{m-1}, \ldots}$ denotes the matrix located on the intersection of the rows $j_m, j_{m-1}, \ldots$ and columns $i_m, i_{m-1}, \ldots$ of the matrix A.

Proof.   An immediate consequence of (4.29) and standard calculus of exterior algebra. ∎

4.4.   Representation of $\mathfrak{a}_\infty$ in F

Matrices in $\overline{\mathfrak{a}}_\infty$ have a finite number of nonzero diagonals and so are finite linear combinations of matrices of the form

$$a_k = \sum_{i \in \mathbb{Z}} \lambda_i E_{i, i+k}, \qquad (4.49)$$

where the λ_i are arbitrary complex numbers. If we try to apply (4.30) to represent a_k in F, we find that for $k \neq 0$, $r(a_k)\psi_m$ is a finite linear combination of semi-infinite monomials in $F^{(m)}$, since the terms appearing in (4.49) vanish if $i + k > m$ or $i \leqslant m$. However, for $k = 0$ we get

$$r(a_0)\psi_m = (\lambda_m + \lambda_{m-1} + \ldots)\psi_m \qquad (4.50)$$

and the sum on the right-hand side can diverge. Since all vectors in $F^{(m)}$ are finite linear combinations of finite excitations of the vacuum vector ψ_m, we conclude that $r(a_k)$ can be defined by (4.30) for $k \neq 0$, but that this definition does not make sense for $k = 0$. We remove the "anomaly" of (4.50) by defining $\hat{r}_m$ by:

$$\hat{r}_m(E_{ij}) = r_m(E_{ij}) \text{ if } i \neq j \text{ or } i = j > 0, \qquad (4.51a)$$

$$\hat{r}_m(E_{ii}) = r_m(E_{ii}) - I \text{ if } i \leqslant 0. \qquad (4.51b)$$

Replacing r_m by $\hat{r}_m$ in (4.50) we see that the right-hand side is now a finite sum:

$$\hat{r}_m(a_0)\psi_m = \left(\sum_{i=1}^{m}\lambda_i\right)\psi_m \text{ if } m \geqslant 1,$$

$$= -\left(\sum_{i=0}^{m+1}\lambda_i\right)\psi_m \text{ if } m \leqslant -1 \text{ and } = 0 \text{ if } m = 0.$$

If $A \in \bar{\mathfrak{a}}_\infty$ then $\hat{r}_m(A)$ maps $F^{(m)}$ into itself. However, while the $r_m(E_{ij})$ obey the commutation rules (4.4), this is no longer the case for $\hat{r}_m(E_{ij})$. We can rewrite (4.4) as a set of four relations:

(i) $[E_{ij}, E_{k\ell}] = 0$ for $j \neq k$, $\ell \neq i$

(ii) $[E_{ij}, E_{j\ell}] = E_{i\ell}$ for $\ell \neq i$

(iii) $[E_{ij}, E_{ki}] = -E_{kj}$ for $j \neq k$

(iv) $[E_{ij}, E_{ji}] = E_{ii} - E_{jj}$. $\hspace{2cm}$ (4.52)

These relations are satisfied by $r_m(E_{ij})$. Since the presence of I in the commutators on the left-hand side of (4.52) will leave the left-hand side unchanged, it follows from (4.51) that the $\hat{r}_m(E_{ij})$ will satisfy the first three equations of (4.52). In the last equation we get

$$[\hat{r}_m(E_{ij}), \hat{r}_m(E_{ji})] = \hat{r}_m(E_{ii}) - \hat{r}_m(E_{jj}) + \alpha(E_{ij}, E_{ji})I$$

where

$$\alpha(E_{ij}, E_{ji}) = -\alpha(E_{ji}, E_{ij}) = 1 \text{ if } i \leqslant 0, j \geqslant 1,$$
$$\alpha(E_{ij}, E_{mn}) = 0 \text{ in all other cases.} \hspace{1cm} (4.53)$$

Thus

$$\hat{r}_m([E_{ij}, E_{k\ell}]) = [\hat{r}_m(E_{ij}), \hat{r}_m(E_{k\ell})] - \alpha(E_{ij}, E_{k\ell}). \hspace{1cm} (4.54)$$

Extending $\hat{r}_m$ to $\bar{\mathfrak{a}}_\infty$ by linearity, we get a projective representation of $\bar{\mathfrak{a}}_\infty$ due to the presence of the scalar summand in (4.54). This can be made into a linear representation of the central extension of $\bar{\mathfrak{a}}_\infty$ *viz.* the Lie algebra $\mathfrak{a}_\infty$ defined by

$$\mathfrak{a}_\infty = \bar{\mathfrak{a}}_\infty \oplus \mathbb{C}c \hspace{2cm} (4.55)$$

with $\mathbb{C}c$ in the center and the bracket

$$[a, b] = ab - ba + \alpha(a, b)c, \qquad (4.56)$$

where the *two-cocycle* $\alpha(a, b)$ is linear in each variable and defined on the E_{ij} by (4.53). Extending $\hat{r}_m$ from $\bar{\mathfrak{a}}_\infty$ to $\mathfrak{a}_\infty$ by $\hat{r}_m(c) = 1$, we obtain a linear representation $\hat{r}_m$ of the Lie algebra $\mathfrak{a}_\infty$ in $F^{(m)}$. It is clear that by extending ω to $\mathfrak{a}_\infty$ by using (4.34) on $\bar{\mathfrak{a}}_\infty$ and by defining $\omega(c) = c$, the representations $\hat{r}_m$ of $\mathfrak{a}_\infty$ are unitary as well. Moreover, one can show in a similar fashion that every unitary irreducible highest weight representation of $g\ell_\infty$ extends to a unitary irreducible representation of $\mathfrak{a}_\infty$.

The algebra $\mathfrak{a}_\infty$ was introduced by Kac and Peterson [1981] and independently by Date, Jimbo, Kashiwara and Miwa [1981].

Let us consider first the shift operators Λ_k which (see (4.9)) form a commutative subalgebra of $\bar{\mathfrak{a}}_\infty$.

Under $\hat{r}_m$ this algebra will become

$$[\hat{r}_m(\Lambda_n), \hat{r}_m(\Lambda_k)] = \alpha(\Lambda_n, \Lambda_k)I.$$

It is a straightforward computation to show that

$$\alpha(\Lambda_n, \Lambda_k) = n\, \delta_{n,-k}, \qquad (4.57a)$$

so that

$$[\hat{r}_m(\Lambda_n), \hat{r}_m(\Lambda_k)] = n\, \delta_{n,-k}. \qquad (4.57b)$$

Note also that

$$\hat{r}_m(\Lambda_0) = mI. \qquad (4.58)$$

Comparing with (2.2) we see that (4.57b) is simply the commutation relations of the oscillator algebra $\mathscr{A}$. Note that the antilinear anti-involution ω of $\mathfrak{a}_\infty$ is consistent with that defined on $\mathscr{A}$ (see Proposition 2.2). Thus we have constructed "fermionic" unitary representations $\hat{r}_m$ of the oscillator algebra.

4.5.   Representations of *Vir* in *F*

We have seen that the algebra $\mathfrak{d}$ can be represented as a two-parameter family of subalgebras of $\bar{\mathfrak{a}}_\infty$ (recall (4.10), (4.11)). Consequently the projective representation of $\mathfrak{d}$ in $F^{(m)}$ under $\hat{r}_m$ must be a linear representation of the 1-dimensional central extension of $\mathfrak{d}$, which we have already determined to be the Virasoro algebra:

$$[\hat{r}(d_i), \hat{r}(d_j)] = (i - j)\hat{r}(d_{i+j}) + \alpha(d_i, d_j).$$

The computation of the two-cocycle is straightforward:

$$\alpha(d_i, d_j) = \sum_{k,\ell}(k - \alpha - \beta(i+1))(\ell - \alpha - \beta(j+1))\alpha(E_{k-i,k}, E_{\ell-j,\ell})$$

$$= \delta_{i,-j}\sum_{k=1}^{i}(k - \alpha - \beta(i+1))(k - i - \alpha + \beta(i-1))$$

$$= \delta_{i,-j}\left(\frac{(i^3 - i)}{12}c_\beta + 2ih_0\right), \tag{4.59}$$

where

$$c_\beta = -12\beta^2 + 12\beta - 2, \quad h_m = \frac{1}{2}(\alpha - m)(\alpha + 2\beta - 1 - m). \tag{4.60}$$

Defining L_i in $F^{(m)}$ by:

$$L_i = \hat{r}(d_i) \text{ if } i \neq 0,$$
$$L_0 = \hat{r}(d_0) + h_0, \tag{4.61}$$

we see that

$$[L_i, L_j] = (i - j)L_{i+j} + \delta_{i,-j}\frac{(i^3 - i)}{12}c_\beta. \tag{4.62}$$

From (4.61) it follows that

$$L_i\psi_m = 0 \text{ for } i > 0; \quad L_0\psi_m = h_m\psi_m. \tag{4.63}$$

We have thus obtained (cf. Feigin and Fuchs [1982]) a representation of the Virasoro algebra on $F^{(m)}$ with central charge c_β and with minimal eigenvalue h_m of the energy operator given by (4.60). Note that these representations of *Vir* are in general non-unitary since the antilinear anti-involution of $\mathfrak{a}_\infty$ is not consistent with that of *Vir*.

Remark 4.2. The following four cases are of special interest:

1) $\beta = 1/2$,
2) $\beta = 0$,
3) $\beta = 1$,
4) $\beta = -1$, $\alpha = 1$.

These are the wedge representation of Vir over the representation of $\mathfrak{d}$ on half-densities, functions, differential forms and vector fields (= the adjoint representation) respectively. In the first case, $c = 1$ and the wedge representation is manifestly unitary (since the underlying representation is unitary); in the second and the third case, $c = -2$; in the fourth case, $c = -26$. The last case is intimately related to the fact that 26 is the critical dimension of the bosonic string theory (see e.g. Feigin [1984], Frenkel-Garland-Zuckerman [1986]). ∎

Remark 4.3. Note that $c_\beta = (6\beta^2 - 6\beta + 1)c_1$. On the other hand, the Chern class of the determinant line bundle λ_β of the vector bundle on the moduli space of algebraic curves, whose fiber over a curve C is the space of differentials of degree β on C, is expressed by the same formula via the Chern class of the Hodge line bundle λ_1 (Mumford's theorem). This coincidence has been explained by Arbarello-De Concini-Kac-Procesi [1987] by establishing a canonical isomorphism between the second cohomology of the Lie algebra of differential operators of degree $\leqslant 1$ and the second singular cohomology of the moduli space of quadruples (C, p, v, L), where C is a smooth genus g Riemann surface, p a point on C, v a nonzero tangent vector to C at p and L a degree $g - 1$ line bundle on C. ∎

LECTURE 5

5.1. Boson-fermion correspondence

In Lecture 4 we realized the oscillator algebra $\mathscr{A}$ as a subalgebra of $\mathfrak{a}_\infty$ by the operators $\hat{r}_m(\Lambda_k)$:

$$[\hat{r}_m(\Lambda_j),\, \hat{r}_m(\Lambda_k)] = j\, \delta_{j,-k}\,. \tag{5.1}$$

For the representation $\hat{r}_m$ on $F^{(m)}$ we have:

$$\hat{r}_m(\Lambda_k)\psi_m = 0 \quad \text{for} \ \ k > 0\,. \tag{5.2}$$

Let us consider all elements of $F^{(m)}$ of the form:

$$\hat{r}_m(\Lambda_{-k_s})\ldots\hat{r}_m(\Lambda_{-k_1})\psi_m \quad (0 < k_1 \leqslant \ldots \leqslant k_s)\,. \tag{5.3}$$

Due to Proposition 2.1, these vectors are linearly independent. All of them with $\sum_i k_i = k$ lie in $F_k^{(m)}$ due to (4.37), and they form a basis of $F_k^{(m)}$ since the dimension of $F_k^{(m)}$ is exactly $p(k)$ (Corollary 4.1).

We have thus obtained an irreducible representation of the algebra of bosonic oscillators $\mathscr{A}$ in the fermionic space $F^{(m)}$ which is isomorphic to the representation of $\mathscr{A}$ in the space B of polynomials in infinitely many variables described in Lecture 2 (see Proposition 2.1). Let σ_m denote this isomorphism:

$$\sigma_m : F^{(m)} \stackrel{\sim}{\to} B^{(m)} = \mathbb{C}[x_1, x_2, \ldots]\,. \tag{5.4}$$

Note that $\sigma_m (\mathbb{C}\psi_m) = \mathbb{C}$ (as these are the elements killed by all Λ_k, $k > 0$). We normalize σ_m by the condition

$$\psi_m \to 1\,. \tag{5.5a}$$

45

For the transported representation $\hat{r}_m^B = \sigma_m\,\hat{r}_m\sigma_m^{-1}$ of $\mathscr{A}$ on $B^{(m)}$ we have $(k > 0)$:

$$\hat{r}_m^B(\Lambda_k) = \frac{\partial}{\partial x_k}\,,$$

$$\hat{r}_m^B(\Lambda_{-k}) = k\,x_k\,,\quad \hat{r}_m^B(\Lambda_0) = m\,. \tag{5.5b}$$

We have put the label 'm' on B in (5.4) to indicate that it is the copy of B corresponding to $F^{(m)}$ for $m \in \mathbb{Z}$. Note that (5.5a and b) completely determine σ_m.

To the principal gradation of $F^{(m)}$ by subspaces $F_k^{(m)}$ of energy k, there corresponds the principal gradation of

$$B^{(m)} = \underset{k\in\mathbb{Z}_+}{\oplus}\,B_k^{(m)}$$

defined by

$$\deg(x_j) = j\,. \tag{5.6}$$

This follows from (4.39).

We have also the contravariant Hermitian form on $B^{(m)}$ transported from $F^{(m)}$ via σ_m which satisfies

$$\langle 1\,|\,1\rangle = 1 \quad\text{and}\quad \hat{r}_m^B(\Lambda_k)^\dagger = \hat{r}_m^B(\Lambda_{-k})\,. \tag{5.7}$$

Note that (2.6) can be rewritten as follows:

$$\langle P\,|\,Q\rangle = \overline{P}\left(\frac{\partial}{\partial x_1},\,\frac{1}{2}\frac{\partial}{\partial x_2},\,\frac{1}{3}\frac{\partial}{\partial x_3},\dots\right)Q(x)\Bigg|_{x=0}\,. \tag{5.8}$$

Here $\overline{P}$ means taking complex conjugates of all coefficients of the polynomial P. Due to Proposition 2.2, this is the transported Hermitian form.

There are two natural questions concerning the isomorphism σ_m, which can be regarded as an algebraic version of the *boson-fermion correspondence*:

1. The semi-infinite monomials $v_{i_m} \wedge v_{i_{m-1}} \wedge \dots$ of $F^{(m)}$ are mapped by σ_m to some polynomials in $B^{(m)}$. What are these polynomials?
2. How can the representation of the subalgebra $\mathscr{A}$ of $\mathfrak{a}_\infty$ in $B^{(m)}$ be extended to the whole Lie algebra $\mathfrak{a}_\infty$?

We defer the answer to the first question to Lecture 6 and take up the second question.

It turns out simpler to deal with F itself rather than with each $F^{(m)}$. Hence we define the direct sum of maps

$$\sigma = \bigoplus_{m \in \mathbb{Z}} \sigma_m , \qquad (5.9\text{a})$$

so that

$$\sigma : F = \bigoplus_{m \in \mathbb{Z}} F^{(m)} \to B \equiv \bigoplus_{m \in \mathbb{Z}} B^{(m)} . \qquad (5.9\text{b})$$

To keep track of the index m on the right-hand side of (5.9b), we introduce a new variable z and put

$$B^{(m)} = z^m \mathbb{C}[x_1, x_2, \ldots] . \qquad (5.10)$$

Thus

$$\sigma(\psi_m) = z^m \qquad (5.11)$$

and we can view B as the polynomial algebra in $x_1, x_2, \ldots$ and z, z^{-1}:

$$B = \mathbb{C}[x_1, x_2, \ldots ; z, z^{-1}] . \qquad (5.12)$$

We let $r^B = \sigma r \sigma^{-1}$ (respectively $\hat{r}^B = \sigma \hat{r} \sigma^{-1}$) be the transported representation of gl_∞ (respectively $\mathfrak{a}_\infty$) from F to B.

5.2. Wedging and contracting operators

We proceed to define wedging and contracting operators in F. Recall from (4.1) that

$$V = \bigoplus_{j \in \mathbb{Z}} \mathbb{C} v_j .$$

Defining the linear functional v_j^* on V by

$$v_j^*(v_i) = \delta_{ij} \quad (i, j \in \mathbb{Z}) , \qquad (5.13)$$

we can define the *restricted dual* of V:

$$V^* = \bigoplus_{i \in \mathbb{Z}} \mathbb{C} v_i^* . \qquad (5.14)$$

Vectors in V and V^* define operators on F as follows. Each $v \in V$ defines a *wedging operator* $\hat{v}$ on F by:

$$\hat{v}(v_{i_1} \wedge v_{i_2} \wedge \ldots) = v \wedge v_{i_1} \wedge v_{i_2} \wedge \ldots . \tag{5.15}$$

Each f in V^* defines a *contracting operator* $\check{f}$ on F by:

$$\check{f}(v_{i_1} \wedge v_{i_2} \wedge \ldots) = f(v_{i_1})v_{i_2} \wedge v_{i_3} \wedge v_{i_4} \wedge \ldots$$
$$- f(v_{i_2})v_{i_1} \wedge v_{i_3} \wedge v_{i_4} \wedge \ldots$$
$$+ f(v_{i_3})v_{i_1} \wedge v_{i_2} \wedge v_{i_4} \wedge \ldots - \ldots . \tag{5.16}$$

Note that the operators $\hat{v}_i$ and $\check{v}_i^*$ are adjoint with respect to the contravariant Hermitian form $\langle \cdot | \cdot \rangle$. The operator $\hat{v}$ maps $F^{(m)}$ into $F^{(m+1)}$, while $\check{f}$ maps $F^{(m)}$ into $F^{(m-1)}$. It is easy to see from (4.32) that

$$r(E_{ij}) = \hat{v}_i \check{v}_j^* . \tag{5.17}$$

Thus

$$\hat{r}(\Lambda_k) = \sum_{i \in \mathbb{Z}} \hat{v}_i \check{v}_{i+k}^* \quad \text{for} \quad k \neq 0 , \tag{5.18a}$$

$$\hat{r}(\Lambda_0) = \sum_{i>0} \hat{v}_i \check{v}_i^* - \sum_{i \leqslant 0} \check{v}_i^* \hat{v}_i . \tag{5.18b}$$

The operators $\{\hat{v}_i, \check{v}_j^* \mid i, j \in \mathbb{Z}\}$ generate a Clifford algebra:

$$[\hat{v}_i, \hat{v}_j]_+ = 0 , \quad [\check{v}_i^*, \check{v}_j^*]_+ = 0 , \quad [\hat{v}_i, \check{v}_j^*]_+ = \delta_{ij} . \tag{5.19}$$

(Here $[\ ,\ ]_+$ stands for the anticommutator: $[a, b]_+ = ab + ba$.) From (5.18a) and (5.19) it is straightforward to verify the following commutation relations, which hold for $j \neq 0$:

$$[\hat{r}(\Lambda_j), \hat{v}_k] = \hat{v}_{k-j} \tag{5.20a}$$

$$[\hat{r}(\Lambda_j), \check{v}_k^*] = -\check{v}_{k+j}^* . \tag{5.20b}$$

Our aim is to determine $r^B(E_{ij})$. From (5.17) it is clear that this can be achieved by transforming $\hat{v}_i$ and $\check{v}_j^*$ by σ.

Remark 5.1. Putting $|0\rangle = \psi_0$ and noting that $\hat{v}_j|0\rangle = 0$ for $j \leqslant 0$ and $\check{v}_j^*|0\rangle = 0$ for $j > 0$, we obtain an isomorphism between the wedge representation of the Clifford algebra and its spin representation, which is more familiar to physicists. ■

5.3. Vertex operators. The first part of the boson-fermion correspondence

Let us introduce the *generating series*

$$X(u) = \sum_{i \in \mathbb{Z}} u^j \hat{v}_j, \ X^*(u) = \sum_{j \in \mathbb{Z}} u^{-j} \hat{v}_j^*, \tag{5.21}$$

where u is a nonzero complex number. As we shall see, the introduction of the generating series for $\hat{v}_i, \hat{v}_j^*$ simplifies the determination of their transforms under σ. Since $X(u)$ is defined by an infinite series, it maps each $F^{(m)}$ into the *formal completion* $\hat{F}^{(m+1)}$ of $F^{(m+1)}$ in which infinite sums of semi-infinite monomials are permitted. Similarly, $X^*(u)$ maps $F^{(m)}$ into $\hat{F}^{(m-1)}$. We define

$$\hat{F} = \bigoplus_{m \in \mathbb{Z}} \hat{F}^{(m)}.$$

The transported operators $\sigma X(u)\sigma^{-1}$ and $\sigma X^*(u)\sigma^{-1}$ map B into $\hat{B}$, where $\hat{B}$ is the space of formal power series in $x_1, x_2, \ldots$ and z, z^{-1}, which are polynomial in z and z^{-1}.

From (5.20a, b) we find that for $j \neq 0$:

$$[\hat{r}(\Lambda_j), X(u)] = u^j X(u), \tag{5.22a}$$

$$[\hat{r}(\Lambda_j), X^*(u)] = -u^j X^*(u). \tag{5.22b}$$

These equations hold in $\hat{F}$; under the isomorphism $\sigma : \hat{F} \xrightarrow{\sim} \hat{B}$ they will hold in $\hat{B}$ as well. We already know the transform of Λ_j, namely, for $j > 0$ we have:

$$\hat{r}^B(\Lambda_j) = \sigma \hat{r}(\Lambda_j)\sigma^{-1} = \partial/\partial x_j$$

$$\hat{r}^B(\Lambda_{-j}) = \sigma \hat{r}(\Lambda_{-j})\sigma^{-1} = jx_j. \tag{5.23}$$

Defining the *vertex operators* $\Gamma(u)$, $\Gamma^*(u)$ by

$$\Gamma(u) = \sigma X(u)\sigma^{-1}$$

$$\Gamma^*(u) = \sigma X^*(u)\sigma^{-1}$$

we see from (5.22a) that $\Gamma(u)$ satisfies the commutation relations:

$$[\partial/\partial x_j, \Gamma(u)] = u^j \Gamma(u) \qquad (5.24a)$$

$$[x_j, \Gamma(u)] = \frac{u^{-j}}{j}\Gamma(u), \qquad (5.24b)$$

with corresponding equations for $\Gamma^*(u)$ coming from (5.22b). The two commutation relations (5.24a, b) suffice to determine $\Gamma(u)$ (up to a constant); similarly, $\Gamma^*(u)$ is also determined by its corresponding commutation relations:

Proposition 5.1. $\Gamma(u)$ and $\Gamma^*(u)$ have the following form on $\hat{B}^{(m)}$:

$$\left.\Gamma(u)\right|_{\hat{B}^{(m)}} = u^{m+1} z \exp\left(\sum_{j\geq 1} u^j x_j\right) \exp\left(-\sum_{j\geq 1} \frac{u^{-j}}{j}\frac{\partial}{\partial x_j}\right) \qquad (5.25a)$$

$$\left.\Gamma^*(u)\right|_{\hat{B}^{(m)}} = u^{-m} z^{-1} \exp\left(-\sum_{j\geq 1} u^j x_j\right) \exp\left(\sum_{j\geq 1} \frac{u^{-j}}{j}\frac{\partial}{\partial x_j}\right). \qquad (5.25b)$$

Proof.   The factor z has to be present on the right-hand side of (5.25a) since $\Gamma(u)$ maps $\hat{B}^{(m)}$ into $\hat{B}^{(m+1)}$. Now let T_u be the operator on $\hat{B}$ defined by

$$(T_u f)(x_1, x_2, \ldots) = f\left(x_1 + u^{-1}, x_2 + \frac{u^{-2}}{2}, \ldots, x_j + \frac{u^{-j}}{j}, \ldots\right).$$

By Taylor's formula

$$T_u = \exp\left(\sum_{j\geq 1} \frac{u^{-j}}{j}\frac{\partial}{\partial x_j}\right). \qquad (5.26)$$

∎

It is now easy to verify that

$$[x_j, \Gamma(u)T_u] = 0, \qquad (5.27)$$

by using (5.24b) and the simple relation

$$[x_j, T_u] = -\frac{u^{-j}}{j}T_u.$$

From (5.27) we conclude that $\Gamma(u)T_u$ contains no differential part, i.e. that

$$\Gamma(u) = z\, f(x_1, x_2, \ldots) \exp\left(-\sum_{j \geqslant 1} \frac{u^{-j}}{j} \frac{\partial}{\partial x_j}\right),$$

where $f(x_1, x_2, \ldots)$ has to be determined. Using (5.24a) and the relation

$$\left[\frac{\partial}{\partial x_i}, \exp\left(-\sum_{j \geqslant 1} u^j x_j\right)\right] = -u^i \exp\left(-\sum_{j \geqslant 1} u^j x_j\right),$$

we find that

$$\left[\frac{\partial}{\partial x_i}, \exp\left(-\sum_{j \geqslant 1} u^j x_j\right)\Gamma(u)\right] = 0. \qquad (5.28)$$

We conclude from (5.28) that

$$\Gamma(u) = c_m(u) z \exp\left(\sum_{j \geqslant 1} u^j x_j\right) \exp\left(-\sum_{j \geqslant 1} \frac{u^{-j}}{j} \frac{\partial}{\partial x_j}\right).$$

We can determine $c_m(u)$ by noting that the coefficient of the vacuum vector ψ_{m+1} of $\hat{F}^{(m+1)}$ in the expansion of $X(u)\psi_m$ is u^{m+1}. This completes the proof of (5.25a). By a similar argument we get (5.25b). ∎

Define the operator $R(u)\colon \hat{B} \to \hat{B}$ by

$$R(u)f(x, z) = uzf(x, uz). \qquad (5.29)$$

Thus if $f(x, z) = z^m g(x_1, x_2, \ldots)$ then

$$R(u)f(x, z) = u^{m+1} z^{m+1} g(x_1, x_2, \ldots).$$

We can now write down the general form of $\Gamma(u)$ and $\Gamma^*(u)$ (cf. Date-Jimbo-Kashiwara-Miwa [1983] and Kac-Peterson [1986]):

Theorem 5.1.

$$\Gamma(u) = R(u) \exp\left(\sum_{j \geqslant 1} u^j x_j\right) \exp\left(-\sum_{j \geqslant 1} \frac{u^{-j}}{j} \frac{\partial}{\partial x_j}\right) \qquad (5.30a)$$

$$\Gamma^*(u) = R(u)^{-1} \exp\left(-\sum_{j \geqslant 1} u^j x_j\right) \exp\left(\sum_{j \geqslant 1} \frac{u^{-j}}{j} \frac{\partial}{\partial x_j}\right). \qquad (5.30b)$$

■

Remark 5.2. Theorem 5.1 is a discrete counterpart of the Skyrme model (see Skyrme [1971]). The idea of its proof is taken from Kac-Kazhdan-Lepowsky-Wilson [1981]. ■

5.4. Vertex operator representations of gl_∞ and $\mathfrak{a}_\infty$

We can now determine the representation of gl_∞ and $\mathfrak{a}_\infty$ in $B^{(m)}$ via the isomorphism σ_m. The basic element is E_{ij} which is represented in $F^{(m)}$ by $\hat{v}_i \check{v}_j^*$. The preceding section has shown us that the transforms of $\hat{v}_i$ and $\check{v}_j^*$ are very complicated, but that it is easier to deal with their generating functions. We shall therefore consider the generating function

$$\sum_{i,j \in \mathbb{Z}} u^i v^{-j} E_{ij}. \qquad (5.31a)$$

The representation in $\hat{F}$ of this generating function under r is simply

$$X(u) X^*(v). \qquad (5.31b)$$

It is a straightforward computation to show, using Theorem 5.1 and

$$(\exp a\, \partial/\partial x)(\exp bx) = (\exp ab)(\exp bx)(\exp a\, \partial/\partial x),$$

that we have

Proposition 5.2.

$$\sum_{i,j \in \mathbb{Z}} u^i v^{-j} r^B(E_{ij}) \equiv \sigma_m(X(u)X^*(v))\sigma_m^{-1} = \frac{(u/v)^m}{1 - (v/u)}\Gamma(u,v) \qquad (5.32)$$

where $\Gamma(u, v)$ is the following *vertex operator*:

$$\Gamma(u, v) = \exp\left(\sum_{j \geqslant 1}(u^j - v^j)x_j\right) \exp\left(-\sum_{j \geqslant 1}\frac{u^{-j} - v^{-j}}{j}\frac{\partial}{\partial x_j}\right) \quad (5.33)$$

and we have assumed that $|v/u| < 1$. ■

In the case of $\hat{r}^B$ we observe from (4.51) that we must simply subtract

$$\sum_{i \leqslant 0}(u/v)^i = \left(1 - \frac{v}{u}\right)^{-1}$$

from the right-hand side of (5.32), where we have once more assumed that $|v/u| < 1$. Thus we arrive at the following:

Proposition 5.3.

$$\sum_{i,j}u^i v^{-j}\hat{r}^B_m(E_{ij}) = \frac{1}{1 - (v/u)}\left(\left(\frac{u}{v}\right)^m\Gamma(u, v) - 1\right). \quad (5.34)$$

■

To calculate $r^B_m(E_{ij})$ or $\hat{r}^B_m(E_{ij})$ we have to determine the coefficient of $u^i v^{-j}$ on the right-hand sides of (5.32) and (5.34).

This *vertex representation* of $\mathfrak{a}_\infty$ was discovered in the case $m = 0$ by Date-Jimbo-Kashiwara-Miwa [1981].

LECTURE 6

6.1. Schur polynomials

In Lecture 5 we asked for the explicit form of the polynomials in the bosonic Fock space $B = \mathbb{C}[x_1, x_2, \ldots]$ which correspond under σ_m to the semi-infinite monomials of $F^{(m)}$ (recall that they form an orthonormal basis in $F^{(m)}$). To find their image in B we need to first introduce the Schur polynomials.

Definition 6.1. *The elementary Schur polynomials $S_k(x)$ are polynomials belonging to $\mathbb{C}[x_1, x_2, \ldots]$ and are defined by the generating function*

$$\sum_{k \in \mathbb{Z}} S_k(x) z^k = \exp \sum_{k=1}^{\infty} x_k z^k. \tag{6.1}$$

Thus

$$S_k(x) = 0 \ \text{ for } \ k < 0, \ S_0(x) = 1, \tag{6.2a}$$

$$S_k(x) = \sum_{k_1 + 2k_2 + \ldots = k} \frac{x_1^{k_1}}{k_1!} \frac{x_2^{k_2}}{k_2!} \cdots \ \text{ for } \ k > 0. \tag{6.2b}$$

In particular

$$S_1(x) = x_1, \ S_2(x) = x_1^2/2 + x_2,$$

$$S_3(x) = x_1^3/6 + x_1 x_2 + x_3, \tag{6.3}$$

$$S_4(x) = x_1^4/24 + x_2^2/2 + x_1^2 x_2/2 + x_1 x_3 + x_4.$$

The elementary Schur polynomials are related to the complete symmetric functions h_k, where h_k is the sum of all monomials of total degree k in the variables $\epsilon_1, \ldots, \epsilon_N$. The generating function for the h_k is

$$\sum_{k \geqslant 0} h_k \, z^k = \prod_{i=1}^{N} (1 - \epsilon_i z)^{-1} . \tag{6.4}$$

To see the connection with the elementary Schur polynomials, substitute

$$x_j = \frac{\epsilon_1^j + \ldots + \epsilon_N^j}{j} \tag{6.5}$$

in the right-hand side of (6.1). We find that this expression reduces to the right-hand side of (6.4), which means that

$$S_k(x) = h_k(\epsilon_1, \ldots, \epsilon_N) . \tag{6.6}$$

We shall denote the set of all partitions by *Par*. Thus $\lambda \in Par$ is a non-increasing finite sequence of positive integers $\{\lambda_1 \geqslant \lambda_2 \geqslant \ldots \geqslant \lambda_k > 0\}$.

Definition 6.2. To each $\lambda = \{\lambda_1 \geqslant \lambda_2 \geqslant \ldots \geqslant \lambda_k\} \in Par$ we associate the *Schur polynomial* $S_\lambda(x)$ defined by the $k \times k$ determinant

$$S_{\lambda_1, \lambda_2, \ldots}(x) = \begin{vmatrix} S_{\lambda_1} & S_{\lambda_1+1} & S_{\lambda_1+2} \cdots \\ S_{\lambda_2-1} & S_{\lambda_2} & S_{\lambda_2+1} \cdots \\ S_{\lambda_3-2} & S_{\lambda_3-1} & S_{\lambda_3} \cdots \\ \cdots\cdots\cdots\cdots\cdots\cdots\cdots \\ \cdots\cdots\cdots\cdots\cdots\cdots\cdots \end{vmatrix} \tag{6.7}$$

$$= \det(S_{\lambda_i+j-i}(x)) .$$

Remark 6.1. It is well-known that

$$S_\lambda(x) = \mathrm{tr}_{\pi_\lambda} \begin{bmatrix} \epsilon_1 & & \\ & \ddots & \\ & & \ddots \\ & & & \epsilon_N \end{bmatrix}$$

where π_λ is the representation of GL_N corresponding to the partition λ and the ϵ_i are related to the x_j by (6.5) (see e.g. Macdonald [1979]). Formula

(6.6) shows this in the case when π_λ is the k-th symmetric power of the natural representation of GL_N. ∎

We find from (6.3) and (6.7):

$$S_{1,1} = x_1^2/2 - x_2\,,$$

$$S_{2,1} = x_1^3/3 - x_3\,, \tag{6.8}$$

$$S_{2,2} = x_1^4/12 - x_1 x_3 + x_2^2\,.$$

It is clear from (6.2) and (6.7) that, with respect to the principal gradation on B introduced in Lecture 2 (in which $\deg x_j = j$), the Schur polynomial $S_{\lambda_1,\lambda_2,\ldots}(x)$ is a homogeneous polynomial of degree $|\lambda| = \lambda_1 + \lambda_2 + \ldots$.

6.2. The second part of the boson-fermion correspondence

We can now state the second part of the boson-fermion correspondence. For simplicity we state the theorem for $m = 0$ (see Corollary 6.1 for the generalization to all m):

Theorem 6.1.

$$\sigma_0(v_{i_0} \wedge v_{i_{-1}} \wedge \ldots) = S_{i_0, i_{-1}+1, i_{-2}+2,\ldots}(x)\,, \tag{6.9}$$

where $i_0 > i_{-1} > \ldots$ and $i_{-k} = -k$ for k sufficiently large.

Proof. Our strategy will be to compute

$$\sigma_0\{R_0(\exp(y_1 \Lambda_1 + y_2 \Lambda_2 + \ldots)) v_{i_0} \wedge v_{i_{-1}} \wedge \ldots\}$$
$$= R_0^B(\exp(y_1 \Lambda_1 + y_2 \Lambda_2 + \ldots)) P(x) \tag{6.10}$$

where

$$P(x) = \sigma_0(v_{i_0} \wedge v_{i_{-1}} \wedge \ldots)\,. \tag{6.11}$$

We shall obtain the result by comparing the coefficient of the vacuum on the two sides of (6.10), recalling from (5.5a) that $\sigma_0(\psi_0) = 1$.

Before we can proceed any further we must first settle a technical problem, since $\exp(y_1 \Lambda_1 + y_2 \Lambda_2 + \ldots)$ is clearly not in GL_∞ (the argument of the exponential is equally not in $g\ell_\infty$). We must consider instead the larger group $\overline{GL}_\infty$ defined by:

$$\overline{GL}_\infty = \{A = (a_{ij}) \,|\, i, j \in \mathbb{Z}, \quad A \text{ invertible and all but a finite}$$
$$\text{number of the } a_{ij} - \delta_{ij} \text{ with } i < j \text{ are } 0\}. \tag{6.12}$$

Thus matrices in $\overline{GL}_\infty$ have only a finite number of nonzero elements below the principal diagonal and it is evident that matrix multiplication is well defined. The Lie algebra of $\overline{GL}_\infty$ is:

$$\overline{g\ell}_\infty = \{(a_{ij}) \,|\, i, j \in \mathbb{Z}, \quad \text{all but finite number of the}$$
$$a_{ij} \text{ with } i < j \text{ are } 0\}.$$

$\overline{GL}_\infty$ and $\overline{g\ell}_\infty$ act not on V, but on a completion $\overline{V}$ of V defined as

$$\overline{V} = \left\{ \sum_j c_j v_j \,|\, c_j = 0 \text{ for } j \gg 0 \right\}.$$

On the other hand, it is easy to see that the representations R and r extend to representations of $\overline{GL}_\infty$ and $\overline{g\ell}_\infty$ on the same space F constructed from V. In particular the formula (4.48) holds for $R_m(A)$ when $A \in \overline{GL}_\infty$. The exponential map is defined on the whole of $\overline{g\ell}_\infty$ and we have

$$\exp r(a) = R(\exp a) \quad \text{for } a \in \overline{g\ell}_\infty.$$

It is clear that if $a = y_1 \Lambda_1 + y_2 \Lambda_2 + \ldots$, then $a \in \overline{g\ell}_\infty$ and $\exp a \in \overline{GL}_\infty$. Hence from the above discussion it follows that (4.48) can be used for $R(\exp a)$. We can now proceed with the proof.

In the bosonic picture, $r_0(\Lambda_k)$ is represented by $\partial/\partial x_k$ for $k > 0$, so that

$$R_0^B(\exp(y_1 \Lambda_1 + y_2 \Lambda_2 + \ldots)) = \exp \sum_{j \geq 1} y_j \frac{\partial}{\partial x_j}.$$

Now let $F(y)$ denote the coefficient of 1 when this operator is applied to $P(x)$. Then

$$F(y) = \exp \left(\sum_{j \geq 1} y_j \frac{\partial}{\partial x_j} \right) P(x) \Bigg|_{x=0}$$

$$= P(x+y) \Bigg|_{x=0} = P(y),$$

i.e.

$$F(y) = P(y). \tag{6.13}$$

Now, $\exp\left(\sum_{k\geqslant 1} \Lambda_k y_k\right) = \exp\left(\sum_{k\geqslant 1} \Lambda_1^k y_k\right) = \sum_{k\geqslant 0} \Lambda_k S_k(y)$, using (6.1). This latter expression can be regarded as a matrix A with matrix elements

$$A_{mn} = S_{n-m}(y) \quad (m, n \in \mathbb{Z}). \tag{6.14a}$$

Recalling from (6.2) that $S_k(x) = 0$ for $k < 0$, we see that $A \in \overline{GL}_\infty$. Hence (6.10) reduces to:

$$\sigma_0(v_{i_0} \wedge v_{i_{-1}} \wedge \ldots) = \text{coefficient of } \psi_0 \text{ in the expansion of}$$
$$\sigma_0\{R_0(A)(v_{i_0} \wedge v_{i_{-1}} \wedge \ldots)\}.$$

We can read off the required coefficient from (4.48), which gives

$$\det(A_{0,-1,-2,\ldots}^{i_0,i_{-1},i_{-2},\ldots}).$$

This expression is the determinant of the matrix of elements of A at the intersections of rows $0, -1, -2, \ldots$ and columns $i_0, i_{-1} i_{-2}, \ldots$ of A. From (6.7) and (6.14a) this is easily seen to be $S_{i_0,i_{-1}+1,i_{-2}+2,\ldots}(y)$. Thus, we have

$$F(y) = S_{i_0,i_{-1}+1,\ldots}(y). \tag{6.14b}$$

Comparing (6.13) and (6.14b) completes the proof. ∎

Corollary 6.1.

$$\sigma_m(v_{i_m} \wedge v_{i_{m-1}} \wedge \ldots) = S_{i_m-m,i_{m-1}-m+1,\ldots}(x). \tag{6.15}$$

∎

Corollary 6.2. In the course of the proof we have determined the action $R_m^B(A)$ of $A \in \overline{GL}_\infty$ in $B^{(m)}$:

$$R_m^B(A)S_\lambda = \sum_{\mu \in Par} \det(A_{\mu_1+m,\mu_2+m-1,\ldots}^{\lambda_1+m,\lambda_2+m-1,\ldots})S_\mu. \tag{6.16}$$

∎

Corollary 6.3. The Schur polynomials form an orthonormal basis in B with respect to the contravariant Hermitian form $\langle \cdot \mid \cdot \rangle$ (defined by (5.8)), i.e.,

$$\langle S_\lambda \mid S_\mu \rangle = \delta_{\lambda, \mu} . \tag{6.17}$$

∎

6.3. An application: structure of the Virasoro representations for $c = 1$

In Lecture 2 we saw that the Virasoro operators

$$L_k = \frac{\epsilon}{2} a_{k/2}^2 + \sum_{j > -k/2} a_{-j} a_{j+k} , \tag{6.18}$$

where $\epsilon = 0$ for k odd, $\epsilon = 1$ for k even, satisfy the Virasoro algebra relations for $c = 1$. The a_k have a representation in $\mathbb{C}[x_1, x_2, \ldots]$ given by (2.3) with $\mathbb{1} = 1$. By the isomorphism established between $\mathbb{C}[x_1, x_2, \ldots]$ and $F^{(0)}$ we can choose the following representation of the a_k for $k > 0$, $\mu \in \mathbb{R}$:

$$a_k = \sqrt{2}\, \hat{r}_0(\Lambda_k), \quad a_{-k} = \frac{1}{\sqrt{2}} \hat{r}_0(\Lambda_{-k}), \quad a_0 = \mu/\sqrt{2} . \tag{6.19}$$

We have made use of the freedom in choosing the ϵ_n in (2.3).

From (6.18) we find that:

$$L_0^{(\mu)} = \mu^2/4 + \sum_{j \geqslant 1} \hat{r}_0(\Lambda_{-j}) \hat{r}_0(\Lambda_j) \tag{6.20a}$$

$$L_1^{(\mu)} = \mu \hat{r}_0(\Lambda_1) + \sum_{j \geqslant 1} \hat{r}_0(\Lambda_{-j}) \hat{r}_0(\Lambda_{j+1}) \tag{6.20b}$$

$$L_2^{(\mu)} = \mu \hat{r}_0(\Lambda_2) + \hat{r}_0(\Lambda_1)^2 + \sum_{j \geqslant 1} \hat{r}_0(\Lambda_{-j}) \hat{r}_0(\Lambda_{j+2}) . \tag{6.20c}$$

This gives us a representation of the Virasoro algebra for $c = 1$ in $F^{(0)}$.

Recall the Definition 3.3 of a singular vector in a representation of *Vir*. Note that it is an immediate consequence of the Virasoro algebra that v is a singular vector if (3.26) holds for $j = 1$ and $j = 2$.

We saw in Lecture 3 that the oscillator representation of the Virasoro algebra for $c = 1$ is a direct sum of unitary, irreducible highest weight representations. Each such representation is generated from a singular vector, *viz.* a highest weight vector. As we shall see in a moment, for generic values of μ in (6.20) there is only one singular vector, *viz.* the vacuum vector ψ_0 of $F^{(0)}$, and this representation is a Verma representation. However, there are special values of μ for which this is not the case:

Lemma 6.1. Let $\mu = -m \in \mathbb{Z}$ and $k \in \mathbb{Z}_+$ be such that $k + m \geqslant 0$. Consider

$$f_{m,k} = v_{k+m} \wedge v_{k+m-1} \wedge \ldots \wedge v_{m+1} \wedge v_{-k} \wedge v_{-k-1} \wedge \ldots \in F^{(0)}. \quad (6.21)$$

Then

$$L_j^{(-m)} f_{m,k} = 0 \quad \text{for} \quad j > 0, \quad (6.22)$$

$$L_0^{(-m)} f_{m,k} = \frac{1}{4}(m + 2k)^2 f_{m,k}. \quad (6.23)$$

Proof. A direct and lengthy computation which we omit. ∎

In Lemma 6.1 we have identified for each $\mu = -m \in \mathbb{Z}$ an infinite sequence of singular vectors. There are two obvious questions. (i) Are there nontrivial singular vectors for other values of μ? (ii) Are there any singular vectors for $\mu = -m$ other than those listed in Lemma 6.1?

Proposition 6.1. (a) If $\mu \notin \mathbb{Z}$ then all singular vectors of *Vir* in $F^{(0)}$ are multiples of the vacuum vector ψ_0.
(b) If $\mu = -m \in \mathbb{Z}$, then any singular vector of *Vir* in $F^{(0)}$ is a linear combination of the $f_{m,k}$ with k, $k + m \in \mathbb{Z}_+$.

Proof. (a) This is an immediate consequence of the Kac determinant formula, which we shall discuss in Lecture 8. (Thus, $M(1, h)$ is irreducible for $h \neq m^2/4$ $(m \in \mathbb{Z})$, i.e. if $\mu \notin \mathbb{Z}$.)
(b) We shall not give the proof. A quick way to see the correctness of the result is as follows. Any singular vector is a linear combination of singular eigenvectors of $L_0^{(-m)}$. We see from (6.20a) that a singular eigenvector has a $L_0^{(-m)}$-eigenvalue of the form $h = m^2/4 + n$, where $n \in \mathbb{Z}_+$ and generates a subrepresentation $V(1, h)$. From the Kac determinant formula it follows that h is of the form $h = (m + j)^2/4$, $j \in \mathbb{Z}_+$. Comparing we see that j is even, i.e. $j = 2k$ and $n = k(k + m)$. Since $n, k \in \mathbb{Z}_+$, we have $k + m \in \mathbb{Z}_+$. These are the sole subrepresentations allowed and from Lemma 6.1 we see

that we have a singular vector for each of them. Hence the space $F^{(0)}$ is a direct sum of irreducible representations $V(1, (m+2k)^2/4)$ for k, $k+m \in \mathbb{Z}_+$ if each representation occurs only once. For the multiplicity question we may appeal to the results of Feigin and Fuchs [1983b] who have shown that the multiplicity is indeed 1. ∎

From Proposition 6.1(b) it follows that:

$$F^{(0)} = \bigoplus_{\substack{k \geqslant 0 \\ k+m \geqslant 0}} V\left(1, \frac{1}{4}(m+2k)^2\right). \tag{6.24}$$

Recall that (see 4.21))

$$\dim_q F^{(0)} = \varphi(q)^{-1}.$$

The subspace $V(1, 1/4(m+2k)^2)$ is generated from $f_{m,k}$ which has degree $k^2 + mk$. Hence from (6.24) we obtain

$$\frac{1}{\varphi(q)} = \sum_{k \in \mathbb{Z}_+} \text{ch}\, V\left(1, \frac{1}{4}(m+2k)^2\right) q^{k^2+mk}. \tag{6.25}$$

Now $f_{m,0} = \psi_0$ so that $V(1, m^2/4)$ is generated from ψ_0. Its character will be lowered from the Verma representation value of $1/\varphi(q)$ by the presence of the subrepresentation $V(1, (m+2)^2/4)$ generated by the singular vector $f_{m,1}$ of degree $m+1$. Hence,

$$\text{ch}\, V(1, m^2/4) \leqslant (1 - q^{m+1})/\varphi(q). \tag{6.26}$$

Comparing (6.25) and (6.26) we see that consistency requires that equality holds in (6.26).

Using the isomorphism between $F^{(0)}$ and $\mathbb{C}[x_1 x_2, \ldots]$ we can summarize the obtained results as follows:

Theorem 6.2. Consider the representation

$$d_k \to L_k^{(\mu)}, C \to 1$$

of the Virasoro algebra on the space $\mathbb{C}[x_1, x_2, \ldots]$. Then:
(a) If $\mu \notin \mathbb{Z}$, this representation is irreducible and hence is a Verma representation.

(b) Let $\mu = -m \in \mathbb{Z}$. Put

$$P_{m,k}(x) = S_{\underbrace{k+m,\ldots,k+m}_{k \text{ times}}}(x) \; (k \geqslant 0, k+m \geqslant 0).$$

Then the $P_{m,k}$ are singular vectors with eigenvalues $(m+2k)^2/4$ and all singular vectors are linear combinations of the $P_{m,k}$. Furthermore, we have:

$$\mathbb{C}[x_1, x_2, \ldots] = \bigoplus_{\substack{k \in \mathbb{Z}_+ \\ k \geqslant -m}} V(1, (m+2k)^2/4).$$

(c) $\operatorname{ch} V(1, m^2/4) = (1 - q^{m+1})/\varphi(q)$. ∎

This theorem is due to several authors: Kac [1979], Segal [1981], Wakimoto-Yamada [1986]. The results of this subsection are not used in the sequel.

LECTURE 7

7.1. Orbit of the vacuum vector under GL_∞

In Lecture 4 we constructed a representation of the group GL_∞ in $F^{(0)}$ and hence in $B = \mathbb{C}[x_1, x_2, \ldots]$ by the boson-fermion correspondence. We shall use this correspondence to study the orbit Ω of the vacuum vector 1 in B under the action of the group GL_∞:

$$\Omega = GL_\infty \cdot 1. \tag{7.1}$$

The set Ω is an infinite-dimensional manifold, each point of which is, as we shall show, a solution of an infinite set of partial differential equations.

We are already familiar with a class of functions contained in Ω:

Proposition 7.1. The Schur polynomials $S_\lambda(x)$ ($\lambda \in Par$) are contained in Ω.

Proof.   By the correspondence between B and $F^{(0)}$, 1 is represented in $F^{(0)}$ by $\psi_0 = v_0 \wedge v_{-1} \wedge \ldots$ and $S_\lambda(x)$ by some $\psi_\lambda = v_{i_0} \wedge v_{i_{-1}} \wedge \ldots$, where $i_{-n} = -n$ for $n \geqslant$ some k. For $A \in GL_\infty$ defined by $Av_{-n} = v_{i_{-n}}$ for $0 \leqslant n \leqslant k - 1$, and $Av_j = v_j$ for all other basis elements v_j, we have: $\psi_\lambda = R_0(A)\psi_0$. Hence for each $\lambda \in Par$, ψ_λ is in $GL_\infty \cdot \psi_0$, i.e. $S_\lambda \in \Omega$.   ∎

We shall use the symbol Ω to denote the orbit of the vacuum vector under GL_∞ interchangeably in B or in $F^{(0)}$.   ∎

7.2. Defining equations for Ω in $F^{(0)}$

Proposition 7.2. If $\tau \in \Omega$, then τ is a solution of the equation

65

$$\sum_{j \in \mathbb{Z}} \hat{v}_j(\tau) \otimes \check{v}_j^*(\tau) = 0. \tag{7.2}$$

Conversely, if $\tau \in F^{(0)}$, $\tau \neq 0$ and τ satisfies (7.2), then $\tau \in \Omega$.

Proof. $\hat{v}_j(\psi_0) = 0$ for $j \leqslant 0$ and $\check{v}_j^*(\psi_0) = 0$ for $j > 0$ so that

$$\sum_{j \in \mathbb{Z}} \hat{v}_j(\psi_0) \otimes \check{v}_j^*(\psi_0) = 0, \tag{7.3}$$

i.e. ψ_0 is a solution of (7.2). Any $\tau \in \Omega$ is of the form

$$\tau = R_0(A)\psi_0, \tag{7.4}$$

where $A \in GL_\infty$. From their definitions (5.15) and (5.16) we easily see that the wedging and contracting operators have the following transformation properties under $R_0(A)$:

$$R_0(A)\hat{v}\, R_0(A)^{-1} = \hat{w}, \quad \text{where } w = Av, \tag{7.5a}$$

$$R_0(A)\check{f}\, R_0(A)^{-1} = \check{g}, \quad \text{where } g = {}^t A^{-1} f, \tag{7.5b}$$

and ${}^t A^{-1}$ is the transpose of A^{-1}. We denote the matrix elements of A and A^{-1} in the basis $\{v_i \,|\, i \in \mathbb{Z}\}$ by a_{ij} and $\bar{a}_{ij}$ respectively, so that:

$$Av_j = \sum_i a_{ji} v_i; \quad {}^t A^{-1} v_j^* = \sum_k \bar{a}_{kj} v_k^*; \quad \sum_j \bar{a}_{kj} a_{ji} = \delta_{ki}. \tag{7.6}$$

If we apply $R_0(A)$ to (7.3) it will act on each component of the tensor product. Using (7.4) we get:

$$\sum_j R_0(A)\hat{v}_j R_0(A)^{-1}(\tau) \otimes R_0(A)\check{v}_j^* R_0(A)^{-1}(\tau) = 0.$$

Using (7.5) and (7.6) this becomes:

$$\sum_{i,j,k} a_{ji} \hat{v}_i(\tau) \otimes \bar{a}_{kj} \check{v}_k^*(\tau) = 0$$

which can be rewritten as:

$$\sum_{i,k} \left(\sum_j \bar{a}_{kj} a_{ji} \right) \hat{v}_i(\tau) \otimes \check{v}_k^*(\tau) = 0$$

i.e., as (7.2).

Thus, any element of Ω satisfies equation (7.2). In particular, as has been pointed out in the proof of Proposition 7.1, any semi-infinite monomial lies in Ω, and hence satisfies equation (7.2).

Conversely, let $\tau \in F^{(0)}$, $\tau \neq 0$ and τ satisfy (7.2). We can write $\tau = \sum_{k=0}^{N} c_k \tau_k$, a linear combination with nonzero coefficients c_k of some semi-infinite monomials τ_k, such that τ_0 is a semi-infinite monomial of greatest (principal) degree; we may assume that $c_0 = 1$. If among the τ_i with $i \geq 1$ there exists a semi-infinite monomial, say τ_1, of the form $r_0(E_{ij})\tau_0$ with $i < j$, we can kill off the term $c_1 \tau_1$ by replacing τ by $R_0(\exp -c_1 E_{ij})\tau$, which again satisfies (7.2) (as shown above). Repeating this procedure several (but a finite) number of times we arrive at an element of the form $\tau' = \tau_0 + b_1 \tau_1 + \cdots + b_s \tau_s$, where b_i are nonzero constants and $\tau_1, \ldots, \tau_s$ are non-proportional semi-infinite monomials that have smaller principal degree than τ_0 and differ from τ_0 by at least two factors. Let

$$T_{ij} = \sum_{n \in \mathbb{Z}} (\hat{v}_n(\tau_i) \otimes \check{v}_n^*(\tau_j) + \hat{v}_n(\tau_j) \otimes \check{v}_n^*(\tau_i)), \quad i, j = 0, 1 \ldots, s.$$

Since τ' satisfies equation (7.2) and $T_{ii} = 0$ (because all τ_i satisfy (7.2)), we obtain

$$\sum_{i=1}^{s} b_i T_{0i} + \sum_{\substack{i,j=1 \\ i<j}}^{s} b_i b_j T_{ij} = 0.$$

Since τ_0 has greater principal degree than $\tau_1, \ldots, \tau_s$, it follows that each sum in the left-hand side is zero and all T_{0i}, $i = 1, \ldots, s$, are linearly independent. Hence $T_{0i} = 0$ for all $i = 1, \ldots, s$. But the equality $T_{0i} = 0$ means:

$$\sum_{n \in \mathbb{Z}} (\hat{v}_n(\tau_0) \otimes \check{v}_n^*(\tau_i) + \hat{v}_n(\tau_i) \otimes \check{v}_n^*(\tau_0)) = 1.$$

If the first term in this equation does not vanish for $n = p$, it must cancel with the second term for some $n = q$. This is only possible if $\check{v}_p^*(\tau_i) = \pm \check{v}_q^*(\tau_0)$, i.e. if τ_0 and τ_i differ by only one factor. This contradiction shows that $\tau' = \tau_0$ and therefore $\tau' \in \Omega$ and $\tau \in \Omega$. ∎

7.3.   Differential equations for Ω in $\mathbb{C}[x_1, x_2, \ldots]$

Consider the expression

$$X(u)\tau \otimes X^*(u)\tau \tag{7.7a}$$

where $X(u)$, $X^*(u)$ are the generating functions defined in (5.21a, b). Then (7.7a) can be rewritten as:

$$\sum_{i,j} u^{i-j} \hat{v}_i(\tau) \otimes \breve{v}_j^*(\tau), \tag{7.7b}$$

and it follows from Proposition 7.2 that $\tau \in \Omega$ if and only if the "constant term" (the term independent of u) in (7.7b) vanishes. In other words, equation (7.2) in the "bosonic picture", via the isomorphism of $F^{(0)}$ and $\mathbb{C}[x_1, x_2, \ldots]$ discussed in Lecture 5, looks as follows:

$$\mathrm{Res}_u \Gamma(u)\tau(x') \otimes \Gamma^*(u)\tau(x'') = 0. \tag{7.8}$$

In more detail, the isomorphism between $F^{(0)}$ and $\mathbb{C}[x_1, x_2, \ldots]$ extends to an isomorphism between $F^{(0)} \otimes F^{(0)}$ and $\mathbb{C}[x'_1, x'_2, \ldots; x''_1, x''_2, \ldots]$, which is the polynomial ring in $x'_1, x'_2, \ldots, x''_1, x''_2, \ldots$. We can transform (7.7a) to the bosonic representation using the identification established in Proposition 5.1:

$$X(u) \to \Gamma(u) = uz \exp\left(\sum_{j \geq 1} u^j x'_j\right) \exp\left(-\sum_{j \geq 1} \frac{u^{-j}}{j} \frac{\partial}{\partial x'_j}\right),$$

$$X^*(u) \to \Gamma^*(u) = z^{-1} \exp\left(-\sum_{j \geq 1} u^j x''_j\right) \exp\left(\sum_{j \geq 1} \frac{u^{-j}}{j} \frac{\partial}{\partial x''_j}\right).$$

Thus (7.7a) becomes:

$$u \exp\left(\sum_{j \geq 1} u^j (x'_j - x''_j)\right) \exp\left(-\sum_{j \geq 1} \frac{u^{-j}}{j} \left(\frac{\partial}{\partial x'_j} - \frac{\partial}{\partial x''_j}\right)\right) \tau(x')\tau(x'').$$

Defining new variables x, y by:

$$x' = x - y, \quad x'' = x + y \tag{7.9a}$$

so that

$$x' - x'' = -2y, \quad \frac{\partial}{\partial x'} - \frac{\partial}{\partial x''} = -\frac{\partial}{\partial y},\tag{7.9b}$$

we deduce from Proposition 7.2

Proposition 7.3. A nonzero element τ of $\mathbb{C}[x_1, x_2, \ldots]$ is contained in Ω if and only if the coefficient of u^0 vanishes in the expression:

$$u \exp\left(-\sum_{j \geqslant 1} 2u^j y_j\right) \exp\left(\sum_{j \geqslant 1} \frac{u^{-j}}{j} \frac{\partial}{\partial y_j}\right) \tau(x-y)\tau(x+y).\tag{7.10}$$

∎

7.4. Hirota's bilinear equations

Definition 7.1. Given a polynomial $P(x_1, x_2, \ldots)$ depending on a finite number of the x_j $(j = 1, 2, \ldots)$, and two functions f and g, we denote by $Pf \cdot g$ the expression

$$P\left(\frac{\partial}{\partial u_1}, \frac{\partial}{\partial u_2}, \ldots\right)(f(x_1 - u_1, x_2 - u_2, \ldots)g(x_1 + u_1, x_2 + u_2, \ldots))\Big|_{u=0}.\tag{7.11}$$

The equation $Pf \cdot g = 0$ is called a *Hirota bilinear equation*.

To illustrate this notation take $P = x$. Then

$$Pf \cdot g = \frac{\partial}{\partial u}(f(x-u)g(x+u))\Big|_{u=0} = -g(x)\frac{\partial f}{\partial x} + f(x)\frac{\partial g}{\partial x}.$$

Let $P = x^n$. Then from Leibniz's formula we obtain:

$$Pf \cdot g = \sum_{k=0}^n (-1)^k \binom{n}{k} \frac{\partial^k f}{\partial x^k} \frac{\partial^{n-k} g}{\partial x^{n-k}}.$$

Remark 7.1. Note that

$$Pf \cdot f \equiv 0 \quad \text{if and only if} \quad P(x) = -P(-x).$$

∎

We now expand the exponentials in (7.10) with the help of the generating functions (6.1) for elementary Schur polynomials:

$$u\left(\sum_{j \geqslant 0} u^j S_j(-2y)\right)\left(\sum_{j \geqslant 0} u^{-j} S_j(\tilde{\partial}_y)\right)\tau(x-y)\tau(x+y),\tag{7.12}$$

where

$$\tilde{\partial}_y = \left(\frac{\partial}{\partial y_1}, \frac{1}{2} \frac{\partial}{\partial y_2}, \frac{1}{3} \frac{\partial}{\partial y_3}, \dots \right). \tag{7.13}$$

Putting equal to zero the term in (7.12) which is independent of u, we get the system of equations:

$$\sum_{j \geqslant 0} S_j(-2y) S_{j+1}(\tilde{\partial}_y) \tau(x - y) \tau(x + y) = 0. \tag{7.14}$$

Now

$$S_{j+1}(\tilde{\partial}_y) \tau(x - y) \tau(x + y) = S_{j+1}(\tilde{\partial}_u) \tau(x - y - u) \tau(x + y + u) \Big|_{u=0}$$

$$= S_{j+1}(\tilde{\partial}_u) \exp\left(\sum_{s \geqslant 1} y_s \frac{\partial}{\partial u_s} \right) \tau(x-u)\tau(x+u) \Big|_{u=0}$$

using Taylor's formula. This last expression can be written as

$$S_{j+1}(\tilde{x}) \exp\left(\sum_{s \geqslant 1} y_s x_s \right) \tau(x) \cdot \tau(x),$$

where

$$\tilde{x} = \left(x_1, \frac{1}{2} x_2, \frac{1}{3} x_3, \dots \right).$$

We thus arrive at

Theorem 7.1. A nonzero polynomial τ is contained in Ω if and only if τ is a solution of the following system of Hirota bilinear equations:

$$\sum_{j=0}^{\infty} S_j(-2y) S_{j+1}(\tilde{x}) \exp\left(\sum_{s \geqslant 1} y_s x_s \right) \tau(x) \cdot \tau(x) = 0, \tag{7.15}$$

where $y_1, y_2, \dots$ are free parameters.

Proof   follows now immediately from Proposition 7.3.   ∎

Equation (7.15) is due to Kashiwara and Miwa [1981].

7.5. The KP hierarchy

If we expand (7.15) in a multiple Taylor series in the variables $y_1, y_2, \ldots$, then each coefficient of this series must vanish, giving us thereby a nonlinear partial differential equation. Let us take the simple case of determining the coefficient of y_r in this expansion. Expanding the exponential in (7.15) we see that y_r appears exactly once with coefficient x_r. In the expansion of the $S_j(-2y)$, y_r appears only in $S_r(-2y)$ with coefficient -2. Thus collecting the coefficient of y_r, we get the Hirota bilinear equation

$$(x_r x_1 - 2S_{r+1}(\tilde{x}))\tau \cdot \tau = 0. \tag{7.16}$$

With the help of (6.3), we find that

$$x_r x_1 - 2S_{r+1}(\tilde{x}) = -x_2 \quad \text{for} \quad r = 1,$$

$$= -x_1^3/3 - 2x_3/3 \quad \text{for} \quad r = 2,$$

$$= x_1 x_3/3 - x_4/2 - x_2^2/4 - x_1^4/12 - x_1^2 x_2/2 \quad \text{for} \quad r = 3.$$

From Remark 7.1 we can drop all odd monomials. Hence $r = 1, 2$ give trivial equations, while for $r = 3$ the even terms give the Hirota equation

$$P\tau \cdot \tau = 0, \quad \text{where} \quad P = x_1^4 + 3x_2^2 - 4x_1 x_3. \tag{7.17}$$

From (7.11) we can rewrite (7.17) as:

$$\left(\frac{\partial^4}{\partial u_1^4} + 3 \frac{\partial^2}{\partial u_2^2} - 4 \frac{\partial^2}{\partial u_1 \partial u_3} \right) \tau(x+u)\tau(x-u)\bigg|_{u=0} = 0.$$

Putting $x_1 = x, x_2 = y, x_3 = t$ and introducing a new function

$$u(x, y, t) = 2 \frac{\partial^2}{\partial x^2} (\log \tau),$$

we find after a calculation that (7.17) becomes the Kadomtzev-Petviashvili (KP) equation:

$$\frac{3}{4} \frac{\partial^2 u}{\partial y^2} = \frac{\partial}{\partial x} \left(\frac{\partial u}{\partial t} - \frac{3}{2} u \frac{\partial u}{\partial x} - \frac{1}{4} \frac{\partial^3 u}{\partial x^3} \right). \tag{7.18}$$

Note that the term in brackets on the right-hand side of (7.18) is the KdV equation. Hence, if u is independent of y, the KP equation reduces essentially to the KdV equation.

As an immediate consequence of Proposition 7.1 and Theorem 7.1 we have:

Corollary 7.1. The following functions are rational solutions of the KP equation:

$$2 \frac{\partial^2}{\partial x^2} \left(\log S_\lambda(x, y, t, c_4, c_5, \ldots) \right),$$

where $c_4, c_5, \ldots$ are arbitrary constants. ∎

Remark 7.2. The family of nonlinear equations (7.15), of which the first is the KP equation, is known as the *KP hierarchy*. The idea that the solutions of the KP hierarchy are parametrized by an infinite dimensional homogeneous space is due to Sato [1981], and it was developed by Date-Jimbo-Kashiwara-Miwa [1981]–[1983]. Equation (7.8) is obviously equivalent to the following equation on the wave functions $\varphi(x, u) = \tau(x)^{-1}(\Gamma(u)\tau(x))$ and $\varphi^*(x, u) = \tau(x)^{-1}(\Gamma^*(u)\tau(x))$:

$$\operatorname{Res}_u \varphi(x', u)\varphi^*(x'', u) = 0. \tag{7.19}$$

We can write:

$$\varphi(x, u) = (1 + \varphi_1(x)u^{-1} + \varphi_2(x)u^{-2} + \ldots)e^{x_1 u + x_2 u^2 + \cdots}.$$

Introduce the associated formal pseudodifferential operator

$$P = 1 + \varphi_1(x)\partial^{-1} + \varphi_2(x)\partial^{-2} + \ldots,$$

and let $A = P\partial P^{-1}$, so that A is a formal pseudodifferential operator of the form

$$A = \partial + a_1(x)\partial^{-1} + a_2(x)\partial^{-2} + \ldots.$$

Letting A_n to be the differential part of A^n (i.e. terms, containing non-negative powers of ∂), it is not difficult to show that equation (7.19) is equivalent to

$$\frac{dA}{dt_n} = [A_n, A], \quad n = 1, 2, \ldots. \tag{7.20}$$

Yet another equivalent formulation is the following "zero curvature" equation

$$\frac{\partial A_m}{\partial x_n} - \frac{\partial A_n}{\partial x_m} = [A_n, A_m], \quad m, n = 1, 2, \ldots \tag{7.21}$$

which is the compatibility condition of the following linear system:

$$A\varphi(x, u) = u\varphi(x, u), \quad \frac{\partial \varphi(x, u)}{\partial x_n} = A_n \varphi(x, u). \tag{7.22}$$

The classical KP equations (7.18) is (7.21) with $m = 2$, $n = 3$, and $x_1 = x$, $x_2 = y$, $x_3 = t$, $u = 2a_1$. See Kashiwara-Miwa [1981] or Kac-van de Leur [1993] for details. ∎

Remark 7.3. Denote by UGM (universal Grassmann manifold) the set of all subspaces U of $V = \oplus_{j \in \mathbb{Z}} \mathbb{C} v_j$ such that U contains $\oplus_{j \leqslant -k} \mathbb{C} v_j$ for $k \gg 0$ as a subspace of codimension k. Note that $\Omega = \{u_0 \wedge u_{-1} \wedge \ldots \mid u_{-j} \in V$ for all j, and $u_{-j} = v_{-j}$ for $j \gg 0\}$, so that we can define a bijective map

$$f : \mathbb{P}\Omega \overset{\sim}{\to} UGM$$

by $f(u_0 \wedge u_{-1} \wedge \ldots) = \sum_j \mathbb{C} u_j$ (here $\mathbb{P}\Omega$ stands for projectivisation of Ω). Thus (due to Theorem 7.1) the set of nonzero polynomial solutions (considered up to a constant factor) of the KP hierarchy is parametrized by the UGM. This is the fundamental observation of Sato [1981]. ∎

7.6. *N*-soliton solutions

The Lie algebra $g\ell_\infty$ has a representation in $\mathbb{C}[x_1, x_2, \ldots]$ defined by the vertex operator $\Gamma(u, v)$. Exponentiating an element of $g\ell_\infty$ gives us an element of GL_∞. The following proposition shows that $\exp(a\Gamma(u, v))$ may be thought of as $1 + a\Gamma(u, v)$:

Proposition 7.4. $\Gamma(u, v)^2 \tau = 0$ for 'good' formal power series τ.

Proof. By Taylor's formula,

$$\exp\left(\sum_{i \geqslant 1} \lambda_i \frac{\partial}{\partial x_i}\right) \tau(x_1, x_2, \ldots) = \tau(x_1 + \lambda_1, x_2 + \lambda_2, \ldots).$$

Hence

$$\Gamma(u,v)\tau(\ldots,x_j,\ldots) = \left(\exp\sum_{j\geqslant 1}(u^j - v^j)x_j\right)\tau\left(\ldots,x_j - \frac{u^{-j} - v^{-j}}{j},\ldots\right).$$

Using the well-known commutation relation

$$e^A e^B = e^B e^A e^c,$$

which holds if

$$[A,B] = cI,$$

and the expansion

$$\log(1-z) = -\sum_{j\geqslant 1}\frac{z^j}{j} \quad (|z| < 1)$$

we find, under the assumption $|u|, |v| < \min(|u'|, |v'|)$:

$$\Gamma(u',v')\Gamma(u,v)\tau(\ldots,x_j,\ldots) = \frac{(u'-u)(v'-v)}{(v'-u)(u'-v)}$$

$$\times\exp\left(\sum_{j\geqslant 1}(u^j - v^j + u'^j - v'^j)x_j\right)\tau\left(\ldots,x_j - \frac{u^{-j} - v^{-j} + u'^{-j} - v'^{-j}}{j},\ldots\right).$$

$$(7.23)$$

The expression (7.23) is valid for all $u \neq v'$, $v \neq u'$ by analytic continuation. Taking the limit $u' \to u$, $v' \to v$ in (7.23), we get $\Gamma(u,v)^2\tau = 0$. ∎

Corollary 7.2. The function

$$\tau_{N;a;u,v}(x) = (1 + a_1\Gamma(u_1,v_1))\ldots(1 + a_N\Gamma(u_N,v_N))\cdot 1 \qquad (7.24)$$

is a solution of the KP hierarchy. (It is known as the τ-function of an *N-soliton solution.*)

Proof. The KP hierarchy can be written symbolically as follows:

$$S(\tau \otimes \tau) = 0,$$

where $S = \sum_{j\in\mathbb{Z}}\hat{v}_j \otimes \check{v}_j^*$ is an operator on $F \otimes F$ commuting with the diagonal action of GL_∞ (see the proof of Proposition 7.2). Since $\Gamma(u,v)$

lies in the completion of $g\ell_\infty$ (Proposition 5.2), we have:

$$\Gamma(u,v)S(\tau \otimes \tau) = S(\Gamma(u,v)\tau \otimes \tau + \tau \otimes \Gamma(u,v)\tau). \qquad (7.25)$$

Since

$$2\Gamma(u,v)\tau \otimes \Gamma(u,v)\tau = \Gamma(u,v)^2(\tau \otimes \tau) - \Gamma(u,v)^2\tau \otimes \tau - \tau \otimes \Gamma(u,v)^2\tau = 0$$

by Proposition 7.4, we deduce from (7.25) that

$$S((1+\Gamma(u,v))\tau \otimes (1+\Gamma(u,v))\tau = S(\tau \otimes \tau) + \Gamma(u,v)S(\tau \otimes \tau).$$

This shows that if τ is a solution of the KP hierarchy, then $(1+a\Gamma(u,v))\tau$ is one as well. Since $\tau = 1$ is a solution, the proof is completed. ∎

The 1-soliton solution of the KP equation (7.18) for $a = 1$ is given by

$$u(x,y,t) = 2\frac{\partial^2(\log \tau_{1;1;u,v}(x))}{\partial x^2},$$

where

$$\tau_{1;1;u,v} = (1+\Gamma(u,v)) \cdot 1$$
$$= 1 + \exp((u-v)x + (u^2 - v^2)y + (u^3 - v^3)t + c),$$

and c is a constant. Thus, we obtain that

$$u(x,y,t) = \frac{(u-v)^2}{2} \cdot \frac{1}{[\cosh 1/2((u-v)x + (u^2 - v^2)y + (u^3 - v^3)t + c)]^2}$$

is a 1-soliton solution of the KP equation.

One easily expands (7.24) using (7.23) to write an explicit formula for the N-soliton solutions (see Date-Jimbo-Kashiwara-Miwa [1981]). We have from (7.23):

$$\Gamma(u_1,v_1)\ldots\Gamma(u_N,v_N) \cdot 1 = \prod_{1 \leqslant i < j \leqslant N} \frac{(u_j - u_i)(v_j - v_i)}{(u_j - v_i)(v_j - u_i)}$$

$$\times \exp\left(\sum_{j=1}^{\infty}\sum_{k=1}^{N}(u_k^j - v_k^j)x_j\right).$$

Hence we obtain:

$$\tau_{N;a;u,v}(x) = \sum_{\substack{0 \leqslant r \leqslant N \\ 1 \leqslant j_1 < j_2 < \ldots < j_r \leqslant N}} \prod_{s=1}^{r} a_{j_s} \prod_{1 \leqslant \lambda < \mu \leqslant r} \frac{(u_{j_\lambda} - u_{j_\mu})(v_{j_\lambda} - v_{j_\mu})}{(u_{j_\lambda} - v_{j_\mu})(v_{j_\lambda} - u_{j_\mu})}$$

$$\times \exp \sum_{k \geqslant 1} \sum_{m=1}^{r} (u_{j_m}^k - v_{j_m}^k)x_k,$$

so that $2(\partial^2/\partial x^2)(\log \tau_{N;a;u,v}(x, y, t, c_4, c_5, \ldots))$ is the N-soliton solution of the KP equation.

LECTURE 8

8.1. Degenerate representations and the determinant $\det_n(c, h)$ of the contravariant form

We saw in Lecture 3 that to every pair of real numbers (c, h) there corresponds the Verma representation $M(c, h)$, which carries a contravariant Hermitian form $\langle \cdot | \cdot \rangle$ and is such that any other highest weight representation is a quotient of $M(c, h)$. Quotiening $M(c, h)$ by its unique maximal proper subrepresentation $J(c, h)$ $(= \text{Ker}\langle \cdot | \cdot \rangle)$ we get the unique irreducible representation $V(c, h)$ with highest weight (c, h).

It is a mathematically interesting question to determine when $M(c, h) = V(c, h)$, i.e. when $M(c, h)$ is irreducible. This problem was solved by Kac [1978]. It is clear that the answer can only depend on the highest weight (c, h). We shall see that generically $M(c, h) = V(c, h)$. If $V(c, h) \neq M(c, h)$ we shall say that $V(c, h)$ is a *degenerate representation* of Vir. In a remarkable recent development, the degenerate representations of Vir have acquired a special significance in the study of the critical behavior of two-dimensional statistical mechanical systems (Belavin-Polyakov-Zamolodchikov [1984a, b]). The classification of the degenerate representations of Vir is, therefore, of interest both in mathematics and physics.

From Proposition 3.4(c) we observe that, for $V(c, h)$ to be degenerate, the contravariant Hermitian form $\langle \cdot | \cdot \rangle$ on $M(c, h)$ must have a nontrivial kernel. Vectors of the form (3.12c) form a linearly independent set of vectors which span $M(c, h)$ and in this basis the matrix of the contravariant form $\langle \cdot | \cdot \rangle$ is

$$\left(\langle d_{-i_t} \ldots d_{-i_1}(v) \, | \, d_{-j_s} \ldots d_{-j_1}(v) \rangle \right), \tag{8.1a}$$

where

$$1 \leqslant i_1 \leqslant \ldots \leqslant i_t, \quad 1 \leqslant j_1 \leqslant \ldots \leqslant j_s, \tag{8.1b}$$

and v is the highest weight vector. However, from Proposition 3.4(b) we know that $M(c, h)$ is a direct sum of finite-dimensional eigenspaces of d_0 which are mutually orthogonal with respect to $\langle \cdot | \cdot \rangle$. Hence the matrix of $\langle \cdot | \cdot \rangle$ is a direct sum of finite-dimensional matrices in each eigenspace $M(c, h)_{h+n}$ of d_0 with eigenvalue $h + n$, $n \in \mathbb{Z}_+$. The restriction of $\langle \cdot | \cdot \rangle$ to $M(c, h)_{h+n}$, the *n-th level*, is the $p(n) \times p(n)$ matrix defined by (8.1a, b) with the additional condition that

$$\sum_k i_k = \sum_k j_k = n. \tag{8.1c}$$

Note that the entries of this matrix are polynomials in c and h.

The condition for the representation $V(c, h)$ to be unitary is that the matrix (8.1a, b, c) should be positive semi-definite for each $n \in \mathbb{Z}_+$. It turns out that this gives us an effective means of determining for which highest weights $V(c, h)$ is unitary. We have an even simpler criterion for degeneracy: a necessary and sufficient condition for the degeneracy of $V(c, h)$ is that for some $n \in \mathbb{Z}_+$ the determinant of the matrix defined by (8.1a, b, c) should vanish. We shall denote the determinant of the matrix defined by (8.1a, b, c) by $\det_n(c, h)$.

The first few values of $\det_n(c, h)$ are easily found:

$$\det_0(c, h) = \langle v | v \rangle = 1,$$

$$\det_1(c, h) = \langle d_{-1}v | d_{-1}v \rangle = 2h,$$

$$\det_2(c, h) = \begin{vmatrix} \langle d_{-2}v | d_{-2}v \rangle & \langle d_{-2}v | d_{-1}^2 v \rangle \\ \langle d_{-1}^2 v | d_{-2}v \rangle & \langle d_{-1}^2 v | d_{-1}^2 v \rangle \end{vmatrix}$$

$$= \begin{vmatrix} 4h + c/2 & 6h \\ 6h & 8h^2 + 4h \end{vmatrix}$$

$$= 2h(16h^2 + 2hc - 10h + c). \tag{8.2}$$

A necessary condition for the representation $V(c, h)$ to be unitary is that $\det_n(c, h) \geqslant 0$ for $n = 0, 1, 2, \ldots$. Thus from $n = 1$ we find that $h \geqslant 0$

is a necessary condition, as we already noted in Proposition 3.5, where we saw that $c \geqslant 0$ is also a necessary condition. The $n = 2$ case gives us more precise information: rewriting $16h^2 + 2hc - 10h + c$ as $(4h-1)^2 + (2h+1)(c-1)$ we observe that the region of the c–h plane defined by

$$0 \leqslant c < 1 - (4h-1)^2/(2h+1), \quad h \geqslant 0$$

is a region of non-unitarity (see Figure 8.1). Thus by studying the sign of $\det_n(c, h)$ we can determine regions of the c–h plane where $M(c, h)$ is not unitary. To proceed further we need a general formula for $\det_n(c, h)$.

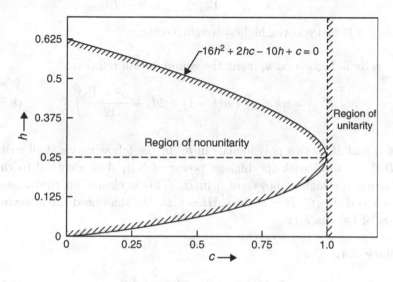

Figure 8.1

8.2. The determinant $\det_n(c, h)$ as a polynomial in h

As the first step in finding a general expression for $\det_n(c, h)$, we shall fix the value of c and consider $\det_n(c, h)$ as a polynomial in h; we shall determine the degree of this polynomial.

Now, h comes from the action of d_0 on the highest weight vector v, and in the Virasoro algebra the commutator $[d_i, d_j]$ gives rise to d_0 only for $i + j = 0$. Examining the matrix (8.1a, b, c) we observe that in each column or row the term giving rise to the maximum number of such commutators is the term lying on the principal diagonal. Thus the leading term in h of

$\det_n(c, h)$ is contained in the expression

$$\prod_{\substack{1 \leqslant i_1 \leqslant \ldots \leqslant i_s \\ \Sigma i_k = n}} \langle d_{-i_s} \ldots d_{-i_1}(v) \mid d_{-i_s} \ldots d_{-i_1}(v) \rangle . \tag{8.3}$$

Lemma 8.1.

$$\langle d^k_{-n}(v) \mid d^k_{-n}(v) \rangle = k! n^k \left(2h + \frac{n^2 - 1}{12} c \right) \left(2h + \frac{n^2 - 1}{12} c + n \right) \ldots$$

$$\times \left(2h + \frac{n^2 - 1}{12} c + n(k - 1) \right) \tag{8.4}$$

where $n, k \in \mathbb{N}$ and v is the highest weight vector.

Proof is by induction on k, using the commutation relation

$$[d_n, d^k_{-n}] = nk \, d^{k-1}_{-n} \left(n(k - 1) + 2d_0 + \frac{n^2 - 1}{12} c \right) . \tag{8.5}$$

∎

Let A and B be two polynomials in h. In the following we shall write $A \sim B$ if the term with the highest power of h in A is identical to the term having the highest power of h in B. This is clearly an equivalence relation and if $A \sim C$, $B \sim D$ then $AB \sim CD$. We shall need the following corollary of Lemma 8.1:

Corollary 8.1.

$$\langle d^k_{-n}(v) \mid d^k_{-n}(v) \rangle \sim k!(2nh)^k . \tag{8.6}$$

∎

Lemma 8.2.

$$\langle d^{j_s}_{-i_s} \ldots d^{j_1}_{-i_1}(v) \mid d^{j_s}_{-i_s} \ldots d^{j_1}_{-i_1}(v) \rangle$$

$$\sim \langle d^{j_s}_{-i_s}(v) \mid d^{j_s}_{-i_s}(v) \rangle \ldots \langle d^{j_1}_{-i_1}(v) \mid d^{j_1}_{-i_1}(v) \rangle ,$$

where $i_1, \ldots, i_s, j_1, \ldots, j_s \in \mathbb{N}$ and $i_1 \neq i_2 \neq \ldots \neq i_s$.

Proof is by induction on $\sum_k j_k$. ∎

If we now apply Lemma 8.2 to (8.3) we get:

Lemma 8.3.

$$\det_n(c, h) \sim \prod_{\substack{r,s \in \mathbb{N} \\ 1 \leqslant rs \leqslant n}} \langle d^s_{-r}(v) \mid d^s_{-r}(v) \rangle^{m(r,s)}$$

where $m(r, s)$ is the number of partitions of n in which r appears exactly s times. ∎

Proposition 8.1. $\det_n(c, h)$ is, for fixed c, a polynomial in h of degree

$$\prod_{\substack{r,s \in \mathbb{N} \\ 1 \leqslant rs \leqslant n}} p(n - rs), \tag{8.7}$$

and the coefficient K of the highest power of h is given by

$$K = \prod_{\substack{r,s \in \mathbb{N} \\ 1 \leqslant rs \leqslant n}} ((2r)^s s!)^{m(r,s)}, \tag{8.8a}$$

where

$$m(r, s) = p(n - rs) - p(n - r(s + 1)). \tag{8.8b}$$

Proof. From Lemma 8.3 and Corollary 8.1 we get the degree of $\det_n(c, h)$ in h to be

$$\sum_{\substack{r,s \in \mathbb{N} \\ 1 \leqslant rs \leqslant n}} s\, m(r, s)$$

with the leading coefficient (8.8a). We thus only have to compute $m(r, s)$, which is the number of partitions of n in which r appears exactly s times. The number of partitions of n in which r appears at least s times is clearly $p(n - rs)$. We subtract from this the number of partitions of n in which r appears at least $s + 1$ times, *viz.* $p(n - r(s + 1))$ to get (8.8b). Then,

$$\sum_{\substack{r,s \in \mathbb{N} \\ 1 \leqslant rs \leqslant n}} s\, m(r, s) = \sum_{1 \leqslant r \leqslant n} \sum_{s=1}^{[n/r]} s(p(n - rs) - p(n - r(s + 1)))$$

$$= \sum_{1 \leqslant r \leqslant n} \sum_{s=1}^{[n/r]} p(n - rs) = \sum_{\substack{r,s \in \mathbb{N} \\ 1 \leqslant rs \leqslant n}} p(n - rs).$$

(The symbol $[n/r]$ means the largest integer not exceeding n/r.) ∎

8.3. The Kac determinant formula

We shall require the following simple lemma in linear algebra:

Lemma 8.4. Let $A(t)$ be a family of linear operators acting in an n-dimensional vector space V and suppose that $A(t)$ is a polynomial function of t. If $A(0)$ has a null space of dimension k, then $\det A(t)$ is divisible by t^k.

Proof. We choose the basis $e_1, \ldots, e_k, e_{k+1}, \ldots, e_n$ in V, such that $e_1, \ldots, e_k$ span the null space of $A(0)$, so that $A(0)e_i = 0$ for $1 \leqslant i \leqslant k$. Then, the first k rows of the matrix of $A(t)$ are divisible by t, proving the lemma. ∎

Lemma 8.5. Consider $\det_n(c, h)$ as a polynomial in h for fixed c. Suppose $\det_n(c, h)$ has a zero at $h = h_0$. Then $\det_n(c, h)$ is divisible by

$$(h - h_0)^{p(n-k)}$$

where k is the smallest positive integer $(1 \leqslant k \leqslant n)$ for which $\det_k(c, h)$ vanishes at $h = h_0$.

Proof. If $\det_n(c, h)$ vanishes at $h = h_0$, then by Proposition 3.4 the matrix of the contravariant form $\langle \cdot \,|\, \cdot \rangle$ on $M(c, h_0)$ has a nonzero kernel when restricted to the n-th level. Thus $M(c, h_0)$ has a nonzero maximal proper subrepresentation $J(c, h_0)$ with a nonzero component $J_n(c, h_0)$ in the n-th level. Let $k \in \mathbb{Z}_+$ be the smallest number such that $J_k(c, h_0) \neq 0$. Picking a nonzero u in $J_k(c, h_0)$, we have:

$$d_0(u) = (h_0 + k)u \quad \text{and} \quad d_n(u) = 0 \quad \text{for } n > 0.$$

Hence u is a singular vector. The application of the universal enveloping algebra $U(Vir)$ to u generates a subrepresentation of $M(c, h_0)$ which is contained in $J(c, h_0)$. The component of this subrepresentation in the n-th level is the linear span of vectors of the form

$$d_{-i_s} \ldots d_{-i_1}(u) \ \left(0 < i_1 \leqslant \ldots \leqslant i_s, \sum_s i_s = n - k \right). \tag{8.9}$$

All such vectors are linearly independent. This follows from the standard fact that a singular vector of a Verma representation generates again a Verma representation. (The latter fact follows from the absence of zero divisors in $U(Vir)$.) The vectors (8.9) thus span a subspace of $J_n(c, h_0)$ of dimension $p(n - k)$. Hence the matrix of the contravariant form restricted to level n has a kernel of dimension at least $p(n - k)$. By Lemma 8.4 it follows that $\det_n(c, h)$ is divisible by $(h - h_0)^{p(n-k)}$. Since u lies in level k, $J(c, h_0)$ has a nontrivial component in level k and $\det_k(c, h)$ must vanish at $h = h_0$. By the definition of u, k is the minimum value for which this happens. ∎

We also require the following lemma, the proof of which we defer to Lecture 12.

Lemma 8.6. Considered as a polynomial in h, $\det_n(c, h)$ has a zero at $h = h_{r,s}(c)$, where

$$
h_{r,s}(c) = \frac{1}{48}[(13 - c)(r^2 + s^2) + \sqrt{(c - 1)(c - 25)}\,(r^2 - s^2)
$$
$$
- 24rs - 2 + 2c], \tag{8.10}
$$

for each pair (r, s) of positive integers such that $1 \leqslant rs \leqslant n$. ∎

Corollary 8.2. $\det_n(c, h)$ is divisible by

$$
\Phi = \prod_{\substack{r,s \in \mathbb{N} \\ 1 \leqslant rs \leqslant n}} (h - h_{r,s}(c))^{p(n-rs)}. \tag{8.11}
$$

Proof. It follows from Lemma 8.6 that $\det_k(c, h)$ has a zero at $h = h_{r,s}(c)$ for $rs \leqslant k \leqslant n$. The corollary now follows from Lemma 8.5. ∎

Theorem 8.1. (Kac [1978]).

$$
\det_n(c, h) = K \prod_{\substack{r,s \in \mathbb{N} \\ 1 \leqslant rs \leqslant n}} (h - h_{r,s}(c))^{p(n-rs)} \tag{8.12}
$$

where $h_{r,s}(c)$ is given by (8.10) and K is the positive constant given by (8.8a, b) (which depends only on n).

Alternatively, let

$$\varphi_{r,r} = h - h_{r,r} = h + (r^2 - 1)(c - 1)/24, \qquad (8.13a)$$

and for $r \neq s$ let

$$\varphi_{r,s} = (h - h_{r,s})(h - h_{s,r}). \qquad (8.13b)$$

Then

$$\det_n(c,h) = K \prod_{\substack{r,s \in \mathbb{N} \\ s \leq r \\ 1 \leq rs \leq n}} \varphi_{r,s}^{p(n-rs)}. \qquad (8.14)$$

Proof. From Corollary 8.2 we know that $\det_n(c,h)$ is divisible by Φ given by (8.11). Moreover the degree of Φ in h agrees with the degree of $\det_n(c,h)$ in h given by (8.7). Hence $\det_n(c,h)$ and Φ can only differ by an overall constant which could depend on c. This constant is the coefficient of the highest power of h which has, however, already been computed in (8.8a, b) and is independent of c. ∎

Corollary 8.3. If $\varphi_{r,s}(c,h) = 0$ and $\varphi_{r',s'}(c,h) \neq 0$ for $r's' < rs$, then $M(c,h)$ has a singular vector of level rs.

Proof. An immediate consequence of the proof of Lemma 8.5 and Theorem 8.1. ∎

Equation (8.14) is very convenient for computations. For $n = 2$, $\det_2(c,h) = 32\varphi_{2,1}\varphi_{1,1}$ which agrees with (8.2).

8.4. Some consequences of the determinant formula for unitarity and degeneracy

Proposition 8.2. (a) The irreducible highest weight representation $V(c,h)$ of the Virasoro algebra is unitary for $c \geq 1$ and $h \geq 0$.
(b) $V(c,h) = M(c,h)$ for $c > 1$, $h > 0$.

Proof. To prove (b), it suffices to show that for each $n \in \mathbb{Z}_+$, $\det_n(c,h) > 0$ for $c > 1$, $h > 0$. Now for $1 \leq r \leq n$, we have $\varphi_{r,r} = h + (r^2 - 1)(c-1)/24 > 0$ for $c > 1$, $h > 0$. For $r \neq s$, $\varphi_{r,s}$ can be rewritten as

$$\varphi_{r,s} = \left(h - \frac{(r-s)^2}{4} \right)^2 + \frac{h}{24}(r^2 + s^2 - 2)(c-1)$$

$$+ \frac{1}{576}(r^2 - 1)(s^2 - 1)(c-1)^2 + \frac{1}{48}(c-1)(r-s)^2(rs+1). \quad (8.15)$$

Thus, $\varphi_{r,s} > 0$ for $1 \leqslant rs \leqslant n, c > 1, h > 0$. Hence from (8.14), $\det_n(c, h) > 0$ for $c > 1, h > 0$, proving (b).

The positivity of the determinants $\det_n(c, h)$ $(n \in \mathbb{Z}_+)$ in the region $c > 1$, $h > 0$ implies that if we can show that the contravariant form is positive definite at a single point in the region, then it is positive definite throughout the region $c > 1$, $h > 0$ and hence positive semidefinite on $M(c, h)$ (so that $V(c, h)$ is unitary) throughout the region $c \geqslant 1, h \geqslant 0$. We have seen, however, in Lecture 3 that we have a manifestly unitary representation of *Vir* in terms of bosonic oscillators for $c = 1, 2, 3, \ldots$ and $h \geqslant 0$, proving (a). ■

Proposition 8.3. (a) $V(1, h) = M(1, h)$ if and only if $h \neq m^2/4 \, (m \in \mathbb{Z})$. (b) $V(0, h) = M(0, h)$ if and only if $h \neq (m^2 - 1)/24 \, (m \in \mathbb{Z})$.

Proof. For $c = 1$ formula (8.12) turns into:

$$\det_n(1, h) = K \prod_{\substack{r,s \in \mathbb{N} \\ 1 \leqslant rs \leqslant n}} \left(h - \frac{(r-s)^2}{4} \right)^{p(n-rs)} \quad (8.16a)$$

so that $\det_n(1, h) \neq 0$ for all $n \in \mathbb{Z}_+$ if and only if $h \neq m^2/4$, $m \in \mathbb{Z}$.

For $c = 0$ formula (8.12) turns into:

$$\det_n(0, h) = K \prod_{\substack{r,s \in \mathbb{N} \\ 1 \leqslant rs \leqslant n}} \left(h - \frac{(3r - 2s)^2 - 1}{24} \right)^{p(n-rs)} \quad (8.16b)$$

so that $\det_n(0, h) \neq 0$ for all $n \in \mathbb{Z}_+$ if and only if $h \neq (m^2 - 1)/24$ $(m \in \mathbb{Z})$. ■

We have shown in Proposition 8.2 (following Kac [1982]) that $V(c, h)$ is unitary in the sector $c \geqslant 1, h \geqslant 0$ of the c–h plane. We know from Proposition 3.5 that a necessary condition for $V(c, h)$ to be unitary is that $c \geqslant 0, h \geqslant 0$. This, therefore, leave the region $0 \leqslant c < 1, h \geqslant 0$ to be discussed.

A simple argument due to Gomes (see Goddard-Olive [1986]) shows that for $c = 0$ the only unitary highest weight representation of *Vir* is the trivial representation in which each d_n is represented by 0. For if the matrix of the contravariant form at level $2N$ is to be positive definite, we require in particular that the matrix

$$\begin{bmatrix} \langle d_{-2N}(v) \mid d_{-2N}(v) \rangle & \langle d_{-N}^2(v) \mid d_{-2N}(v) \rangle \\ \langle d_{-2N}(v) \mid d_{-N}^2(v) \rangle & \langle d_{-N}^2(v) \mid d_{-N}^2(v) \rangle \end{bmatrix}$$

be positive definite. Evaluating the determinant for $c = 0$ we obtain

$$4N^3 h^2 \, (8h - 5N),$$

which is negative for large N unless $h = 0$. If $h = 0$, then $V(0,0)$ is the 1-dimensional trivial representation.

In the region $0 \leqslant c < 1$ it is convenient to use the following parametrization of c:

$$c(m) = 1 - \frac{6}{(m+2)(m+3)}. \tag{8.17}$$

The region $0 \leqslant c < 1$ corresponds to $m \geqslant 0$. This parametrization has the effect of rationalizing the expression (8.10) for $h_{r,s}$:

$$h_{r,s}(m) = \frac{((m+3)r - (m+2)s)^2 - 1}{4(m+2)(m+3)}. \tag{8.18}$$

Now from Figure 8.1 we know that not all points in the region $0 \leqslant c < 1, h > 0$ can correspond to unitary highest weight representations of *Vir*. Considering that the $n = 2$ case alone eliminates a large region, we might well suspect that the infinite number of nonlinear constraints coming from higher levels rule out the entire region. We know, however, from Lecture 3 that $c = 1/2$ and $h = 0, 1/16$ or $1/2$ give rise to nontrivial unitary representations. By a detailed analysis of the Kac determinant formula, Friedan-Qiu-Shenker [1985], [1986] have shown that the only *possible* places of unitarity of $V(c, h)$ in the region $0 \leqslant c < 1, h \geqslant 0$, are the discrete set of points:

$$(c(m), h_{r,s}(m)), \quad \text{where } m, r, s \in \mathbb{Z}_+ \text{ and } 1 \leqslant s \leqslant r \leqslant m+1. \tag{8.19}$$

For $m = 0$ we get $c = h = 0$, which is the trivial representation. For $m = 1$ we get $c = 1/2$ and $h = 0, 1/2, 1/16$ in agreement with our construction in Lecture 3.

We shall show in Lecture 12 that the representations $V(c, h)$ of *Vir* are indeed unitary for every pair (c, h) belonging to the "discrete series" (8.19).

Remark 8.1. Belavin-Polyakov-Zamolodchikov [1984b] pointed out that the series (8.19) correspond to the most important 2-dimensional statistical mechanical models. Then, Friedan-Qiu-Shenker [1985] interpreted this set from the point of view of the unitarity property. ■

Remark 8.2. Another interesting region of irreducibility of the $M(c, h)$ is: $c \leqslant 1, 24h < c - 1$. It is easy to see that in this region one has:

$$\sum_{n \in \mathbb{Z}_+} s_n \, q^n = \prod_{n \geqslant 1} (1 + q^n)^{-1},$$

where s_n denotes the signature of $\langle \cdot \mid \cdot \rangle$ on the n-th level. This is shown by replacing in (3.24), λ and μ by $i\lambda$ and $i\mu$ and taking $\omega(a_n) = -a_{-n}$. The calculation of the series $\sum s_n \, q^n$ for arbitrary $M(c, h)$ is a very interesting problem, which was open at the time of publication of the first edition. This problem has been solved by A. Kent [1991]. ■

LECTURE 9

9.1. Representations of loop algebras in $\bar{a}_\infty$

In Lecture 4 we realized the Lie algebra $\mathfrak{d}$ as a subalgebra of $\bar{a}_\infty$, and, using this, we then constructed highest weight representations in $F^{(m)}$ of its central extension. We shall follow this procedure now for loop algebras.

Definition 9.1. Let $g\ell_n$ denote the Lie algebra of all $n \times n$ matrices with complex entries acting in $\mathbb{C}^n$ and let $\mathbb{C}[t, t^{-1}]$ denote the ring of Laurent polynomials (i.e. polynomials in t and t^{-1}). We define the *loop algebra* $\widetilde{g\ell}_n$ as $g\ell_n(\mathbb{C}[t, t^{-1}])$ i.e. as the complex Lie algebra of $n \times n$ matrices with Laurent polynomials as entries.

Remark 9.1. We can view $\widetilde{g\ell}_n$ as the Lie algebra of maps from the unit circle S^1 to the Lie algebra $g\ell_n$, with finite Fourier series, and the Lie bracket defined pointwise. This accounts for the name 'loop algebra'. ∎

An element of $\widetilde{g\ell}_n$ has the form

$$a(t) = \sum_k t^k a_k \ (a_k \in g\ell_n), \tag{9.1}$$

where k runs over a finite subset of $\mathbb{Z}$. Since the matrices e_{ij} $(1 \leqslant i, j \leqslant n)$, which have 1 as the (i, j) entry and 0 elsewhere, form a basis of $g\ell_n$, it is clear that the matrices

$$e_{ij}(k) \equiv t^k e_{ij} \ (1 \leqslant i, j \leqslant n \ \text{and} \ k \in \mathbb{Z}) \tag{9.2}$$

form a basis of $\widetilde{g\ell}_n$. The elements of $\widetilde{g\ell}_n$ form an associative algebra with multiplication defined on the basis elements by

$$e_{ij}(k)e_{mn}(\ell) = t^{k+\ell} e_{ij}e_{mn} = \delta_{jm} e_{in}(k+\ell). \tag{9.3}$$

The Lie bracket on $\widetilde{g\ell}_n$ is the commutator of the associative multiplication defined by (9.3):

$$[e_{ij}(k), e_{mn}(\ell)] = \delta_{jm} e_{in}(k+\ell) - \delta_{ni} e_{mj}(k+\ell). \tag{9.4}$$

The vector space in which $g\ell_n$ acts is $\mathbb{C}^n$, which has a standard basis $u_1, \ldots, u_n$ of $n \times 1$ column vectors in which u_k ($1 \leqslant k \leqslant n$) has 1 in the k-th row and 0 elsewhere. The loop algebra $\widetilde{g\ell}_n$ acts in $\mathbb{C}[t, t^{-1}]^n$, which consists of $n \times 1$ column vectors with Laurent polynomials in t as entries. The vectors

$$v_{nk+j} = t^{-k} u_j \tag{9.5}$$

form a basis of $\mathbb{C}[t, t^{-1}]^n$ (over $\mathbb{C}$) indexed by $\mathbb{Z}$. Thus we obtain an identification of $\mathbb{C}[t, t^{-1}]^n$ with $\mathbb{C}^\infty$. From (9.2) and (9.5),

$$e_{ij}(k)v_{ns+j} = v_{n(s-k)+i}. \tag{9.6}$$

For $a(t) \in \widetilde{g\ell}_n$ we shall denote the corresponding matrix in $\bar{a}_\infty$ by $\tau(a(t))$. Then from (9.6) we can deduce that $e_{ij}(k)$ has, in the notation of Lecture 4 (see (4.3)), the following matrix representation in $\bar{a}_\infty$:

$$\tau(e_{ij}(k)) = \sum_{s \in \mathbb{Z}} E_{ns+i,n(s+k)+j}. \tag{9.7}$$

More generally, given $a(t) \in \widetilde{g\ell}_n$ as in (9.1), the corresponding matrix in $\bar{a}_\infty$ has the following block form:

$$\tau(a(t)) = \begin{bmatrix} \cdots\cdots\cdots\cdots\cdots\cdots\cdots \\ \cdots a_{-1}\, a_0 \quad a_1 \quad \cdots\cdots\cdots \\ \cdots\cdots\cdots a_{-1}\, a_0 \quad a_1 \cdots\cdots \\ \cdots\cdots\cdots\cdots a_{-1}\, a_0 \cdots\cdots \\ \cdots\cdots\cdots\cdots\cdots\cdots\cdots \end{bmatrix} \tag{9.8}$$

We regard (9.8) as a matrix of $\bar{a}_\infty$ in which the elements on each diagonal parallel to the principal diagonal form a periodic sequence with period n.

We note some properties of the mapping τ in the following proposition:

Proposition 9.1. (a) τ is an injective homomorphism of associative algebras, and hence Lie algebras, since the Lie bracket is the commutator.

(b) The image of $a(t) = \sum_j a_j t^j$ under τ is a strictly upper triangular matrix if and only if

$$a(t) = a_0 + a_1 t + a_2 t^2 + \ldots \quad \text{with } a_0 \text{ strictly upper triangular.} \quad (9.9)$$

(c) The shift operator Λ_j is the image under τ of $(a + tb)^j$, where

$$a = \sum_{i=1}^{n-1} e_{i,i+1}, \quad b = e_{n1}.$$

(d) Let $X(k) = t^k X$ be an element of $\widetilde{g\ell}_n$, where X is in $g\ell_n$. Define an antilinear anti-involution ω on $\widetilde{g\ell}_n$ by

$$\omega(X(k)) = t^{-k} X^\dagger, \quad (9.10)$$

where $X^\dagger$ denotes the Hermitian adjoint of the $n \times n$ matrix X. Then

$$\tau(\omega(X(k))) = (\tau(X(k)))^\dagger, \quad (9.11)$$

where the symbol $\dagger$ on the right-hand side indicates the matrix Hermitian adjoint in $\bar{a}_\infty$.

Proof is straightforward. Let us check, for example, (c). Note that by (a):

$$\tau((a+bt)^j) = \left(\sum_{i=1}^{n-1} \tau(e_{i,i+1}) + \tau(te_{n1}) \right)^j$$

$$= \left(\sum_{i=1}^{n-1} \sum_s E_{ns+i,ns+i+1} + \sum_s E_{ns+n,n(s+1)+1} \right)^j$$

$$= \left(\sum_s \sum_{i=1}^{n} E_{ns+i,ns+i+1} \right)^j = \Lambda_1^j = \Lambda_j. \qquad \blacksquare$$

Remark 9.2. Viewing $\widetilde{g\ell}_n$ as a loop algebra, $\omega(a(t))$ is simply a pointwise Hermitian adjoint of the loop $a(t)$. Thus the corresponding "compact form"

$$\{a(t) \in \widetilde{g\ell}_n \mid \omega(a(t)) = -a(t)\}$$

is simply the Lie algebra of maps (with finite Fourier series) of S^1 into su_n.   ∎

9.2.   Representations of $\hat{g\ell}_n'$ in $F^{(m)}$

In the previous section we have given a realization of $\widetilde{g\ell}_n$ as a subalgebra of $\bar{a}_\infty$. As such it will have a projective representation in the wedge space $F^{(m)}$ (see §4.4) and its central extension

$$\hat{g\ell}_n' = \widetilde{g\ell}_n \oplus \mathbb{C}M \tag{9.12}$$

will have a linear representation as a subalgebra of a_∞, the central extension of $\bar{a}_\infty$. We can compute the two-cocycle α on a pair of basis elements of $\widetilde{g\ell}_n$ of the form (9.2), using the representation (9.7) and (4.53), and we find

$$\alpha(\tau(e_{ij}(k)), \tau(e_{pq}(\ell))) = \delta_{iq}\,\delta_{jp}\,\delta_{\ell+k,0}\,k\,. \tag{9.13}$$

It now follows by linearity that if $X(k) = t^k X, Y(\ell) = t^\ell Y$ then

$$\alpha(\tau(X(k)), \tau(Y(\ell))) = \delta_{k,-\ell}\,k\,\mathrm{tr}(XY)\,, \tag{9.14a}$$

where tr denotes the trace in $g\ell_n$. For general elements $a(t), b(t)$ in $\widetilde{g\ell}_n$ the formula (9.14a) can be written as follows:

$$\alpha(\tau(a(t)), \tau(b(t))) = \mathrm{Res}_0\ \mathrm{tr}\,a'(t)b(t) \tag{9.14b}$$

where $a'(t)$ is the derivative of a with respect to t and Res_0 is the residue at $t = 0$, i.e., the coefficient of $1/t$.

We have thus been led by our search for highest weight representations of $\widetilde{g\ell}_n$ to consider its central extension $\hat{g\ell}_n'$.

Definition 9.2. The Lie algebra $\hat{g\ell}_n'$, defined by (9.12) and the commutation relations

$$[a(t), M] = 0\,,$$
$$[a(t), b(t)] = a(t)b(t) - b(t)a(t) + (\mathrm{Res}_0\ \mathrm{tr}\,a'(t)b(t))M\,, \tag{9.15}$$

is called an *affine Kac-Moody algebra*, or simply an *affine algebra, associated to* $g\ell_n$.

We shall frequently use the commutation relations (9.15) for the elements $X(k) = t^k X, Y(m) = t^m Y$:

$$[X(k), Y(m)] = [X, Y](k+m) + k\delta_{k,-m}(\operatorname{tr} XY)M. \qquad (9.16)$$

9.3. The invariant bilinear form on $\hat{g\ell}_n$. The action of $\widetilde{GL}_n$ on $\hat{g\ell}_n$

Definition 9.3. A bilinear form $(\cdot \,|\, \cdot)$ on a Lie algebra $\mathfrak{g}$ with the Lie bracket $[\cdot, \cdot]$ is *invariant* if

$$([x,y]\,|\,z) = (x\,|\,[y,z]) \quad \text{for all } x, y, z \in \mathfrak{g}. \qquad (9.17)$$

The bilinear form on $g\ell_n$ defined by

$$(X\,|\,Y) = \operatorname{tr}(XY) \qquad (9.18)$$

is symmetric nondegenerate and invariant because of the properties of the trace. Now $g\ell_n$ is the Lie algebra of the group GL_n and $(\cdot \,|\, \cdot)$ has the property of being invariant under the adjoint action Ad of this group:

$$(\operatorname{Ad}(A)(X)\,|\,\operatorname{Ad}(A)(Y)) \equiv (AXA^{-1}\,|\,AYA^{-1}) = (X\,|\,Y) \quad \text{for all } A \in GL_n. \qquad (9.19)$$

(Of course (9.17) is the infinitesimal equivalent for $g\ell_n$ of (9.19).)

We can define a bilinear form on $\widehat{g\ell}_n$ in analogy with (9.18):

$$(X(k)\,|\,Y(m)) = \delta_{k+m,0} \operatorname{tr}(XY). \qquad (9.20a)$$

This definition extends by linearity to general elements $a(t), b(t)$ of $\widetilde{g\ell}_n$ as follows:

$$(a(t)\,|\,b(t)) = \operatorname{Res}_0 \, t^{-1} \operatorname{tr}(a(t)b(t)). \qquad (9.20b)$$

It is easily checked that $(\cdot \,|\, \cdot)$ is a symmetric invariant nondegenerate bilinear form on $\widetilde{g\ell}_n$. It also has the property, which is analogous to (9.19), of being invariant under the adjoint action Ad of the group $\widetilde{GL}_n$, where

$$\widetilde{GL}_n \equiv GL_n(\mathbb{C}[t, t^{-1}]) \qquad (9.21)$$

is the group of all invertible $n \times n$ matrices over $\mathbb{C}[t, t^{-1}]$, *viz.* for all $A(t) \in \widetilde{GL}_n$ we have

$$(A(t)a(t)A^{-1}(t) \mid A(t)b(t)A^{-1}(t)) = (a(t) \mid b(t)). \qquad (9.22)$$

The form $(\cdot \mid \cdot)$ can be extended to $\hat{g\ell}'_n$ simply by defining

$$(M \mid \widetilde{g\ell}_n) = 0, \quad (M \mid M) = 0. \qquad (9.23)$$

This definition preserves all the previous properties, except that now it is, of course, degenerate.

Definition 9.4. It is convenient from several points of view to enlarge $\hat{g\ell}'_n$ by adding one more generator d:

$$\hat{g\ell}_n = \hat{g\ell}'_n \oplus \mathbb{C}d, \qquad (9.24)$$

where the commutation relations with the new generator d are:

$$[d, M] = 0, \quad [d, X(k)] = kX(k), \quad \text{i.e. } [d, a(t)] = t\, a'(t). \qquad (9.25)$$

As before, $a'(t) = da/dt$. The Lie algebra $\hat{g\ell}_n$ is also called an affine (Kac-Moody) algebra.

Proposition 9.2. The affine algebra $\hat{g\ell}_n = \widetilde{g\ell}_n \oplus \mathbb{C}M \oplus \mathbb{C}d$ carries a nondegenerate symmetric invariant bilinear form $(\cdot \mid \cdot)$ defined by

$$\left.\begin{aligned}
(a(t) \mid b(t)) &= \mathrm{Res}_0 \; t^{-1} \, \mathrm{tr}(a(t)b(t)) \quad \text{for } a(t), b(t) \in \widetilde{g\ell}_n, \\
(M \mid a(t)) &= 0, \quad (M \mid M) = 0, \\
(d \mid a(t)) &= 0, \quad (d \mid M) = 1, \quad (d \mid d) = 0.
\end{aligned}\right\} \qquad (9.26)$$

Proof. It is clear that the condition $(d \mid M) = 1$ ensures that $(\cdot \mid \cdot)$ is nondegenerate. Since $[M, d] = 0$, to prove invariance it suffices to show that for all $X(k), Y(m) \in \widetilde{g\ell}_n$,

$$([X(k), d] \mid Y(m)) = (X(k) \mid [d, Y(m)]).$$

This, however, follows immediately from (9.25) and (9.20a). ∎

Since, by (9.25), $dA - Ad = tA'$, we obtain

$$AdA^{-1} = d - tA'A^{-1} \qquad (9.27)$$

for $A \in \widetilde{GL}_n$ acting on $\widetilde{g\ell}_n \oplus \mathbb{C}d$.

Now we want to lift the action of $\widetilde{GL}_n$ from $\widetilde{g\ell}_n \oplus \mathbb{C}d$ to $\hat{g\ell}_n$. It is clear from the commutation relations of $\hat{g\ell}_n$, *viz.* (9.15) and (9.25), that we must have:

$$\mathrm{Ad}(A(t))\,(M) = M\,,$$

$$\mathrm{Ad}(A(t))\,(x(t)) = Ax(t)A^{-1} + \lambda(A, x)M\,,$$

$$\mathrm{Ad}(A(t))(d) = d - tA'A^{-1} + \mu(A)M\,,$$

where $A(t) \in \widetilde{GL}_n, x(t) \in \widetilde{g\ell}_n$ and $\lambda(A, x), \mu(A) \in \mathbb{C}$, and we used (9.27) in the last line. We demand further the $\widetilde{GL}_n$-invariance of the form $(\cdot \mid \cdot)$ on $\hat{g\ell}_n$. This gives us

$$0 = (x \mid d) = (\mathrm{Ad}(A)(x) \mid \mathrm{Ad}(A)(d))$$

$$= (AxA^{-1} + \lambda M \mid d - tA'A^{-1} + \mu M)\,,$$

from which $\lambda = \mathrm{Res}_0 \,\mathrm{tr}\, A'xA^{-1}$. Similarly, from

$$0 = (d \mid d) = (\mathrm{Ad}(A)(d) \mid \mathrm{Ad}(A)(d))$$

we get $\mu = -1/2 \,\mathrm{Res}_0 \,\mathrm{tr}(t(A'A^{-1})^2)$.

Summarizing:

$$\mathrm{Ad}(A(t))(M) = M\,, \qquad (9.28a)$$

$$\mathrm{Ad}(A(t))(x(t)) = Ax(t)A^{-1} + \mathrm{Res}_0 \,\mathrm{tr}(A'x(t)A^{-1})M\,, \qquad (9.28b)$$

$$\mathrm{Ad}(A(t))(d) = d - tA'A^{-1} - \frac{1}{2}\,\mathrm{Res}_0 \,\mathrm{tr}(t(A'A^{-1})^2)M\,. \qquad (9.28c)$$

One checks immediately that these formulas indeed define an automorphism of the Lie algebra $\hat{g\ell}_n$.

9.4.   Reduction from $\mathfrak{a}_\infty$ to $\hat{s\ell}_n$ and the unitarity of highest weight representations of $\hat{s\ell}_n$

We showed in §9.1 (see Proposition 9.1) that $\widetilde{g\ell}_n$ is a subalgebra of $\bar{\mathfrak{a}}_\infty$ and that the antilinear anti-involution ω on $\widetilde{g\ell}_n$ coincides with the one induced from $\bar{\mathfrak{a}}_\infty$. The central extension $\hat{g\ell}_n'$ of $\widetilde{g\ell}_n$ is a subalgebra of $\mathfrak{a}_\infty$ and if we put in addition $\omega(M) = M$, the antilinear anti-involutions on $\hat{g\ell}_n'$ and $\mathfrak{a}_\infty$ are consistent. Moreover, $\hat{g\ell}_n'$ contains the principal subalgebra $\mathscr{A}$ of $\mathfrak{a}_\infty$ (recall Proposition 9.1(c)). We know from Lecture 4 that $\mathfrak{a}_\infty$ has a sequence of fundamental irreducible representations $\hat{r}_m$ in $F^{(m)}$, which remain irreducible when restricted to the subalgebra $\mathscr{A}$ (see Section 5.1). Thus $\hat{g\ell}_n'$ has a sequence of irreducible representations $\hat{\pi}_m$ in $F^{(m)}$ and for $a(t)$ as in (9.9) we have:

$$\hat{\pi}_m(\tau(a(t)))\psi_m = 0. \tag{9.29}$$

Moreover, $\hat{\pi}_m(M)$ acts by 1 in each $F^{(m)}$, while the action of the diagonal elements

$$\hat{\pi}_m(\tau(e_{ii}(0)))\psi_m = \sum_{s \in \mathbb{Z}} \hat{r}_m(E_{ns+i,ns+i})\psi_m \tag{9.30}$$

can be determined from:

$$\hat{r}_m(E_{ns+i,ns+i})\psi_m = \lambda(ns + i, m)\psi_m \tag{9.31}$$

where

$$\lambda(j, m) = \begin{cases} 1 & \text{if } m \geqslant j > 0 \\ -1 & \text{if } 0 \geqslant j > m \\ 0 & \text{otherwise}. \end{cases}$$

It is easy to see that the representations $\hat{\pi}_m$ of $\hat{g\ell}_n'$ extend uniquely to $\hat{g\ell}_n$ (so that (9.25) holds) subject to the condition

$$\pi_m(d)\psi_m = 0. \tag{9.32}$$

The Lie algebra $s\ell_n$ of $n \times n$ complex traceless matrices is a Lie subalgebra of $g\ell_n$. Hence $\widetilde{s\ell}_n$, $\hat{s\ell}_n'$ and $\hat{s\ell}_n$ can be defined and are Lie subalgebras of $\widetilde{g\ell}_n$, $\hat{g\ell}_n'$ and $\hat{g\ell}_n$ respectively. Consequently $\hat{s\ell}_n'$ is also a subalgebra of $\mathfrak{a}_\infty$. The Cartan subalgebra $\mathfrak{h}$ of sl_n has as basis the diagonal traceless matrices $e_{ii} - e_{i+1,i+1}$ $(1 \leqslant i \leqslant n-1)$. The *Cartan subalgebra* $\hat{\mathfrak{h}}$ of $\hat{sl}_n$ is spanned by

$$\{h_i \equiv e_{ii}(0) - e_{i+1,i+1}(0) \;\; (1 \leqslant i \leqslant n-1); \; M; \; d\}.$$

We shall, however, choose as basis

$$\{h_0 \equiv M + e_{n,n}(0) - e_{1,1}(0); \; h_i \;(1 \leqslant i \leqslant n-1); \; d\}.$$

Define linear functionals ω_j $(j = 0, 1, \ldots, n-1)$ on $\hat{\mathfrak{h}}$ by

$$\omega_j(h_i) = \delta_{ij} \;(0 \leqslant i, j \leqslant n-1); \;\; \omega_j(d) = 0.$$

Using (9.30) and (9.31) we can verify that

$$\hat{\pi}_m(h_i)\psi_m = \omega_{m'}(h_i)\psi_m, \tag{9.33a}$$

or, more generally, using (9.32):

$$\hat{\pi}_m(h)\psi_m = \omega_{m'}(h)\psi_m \;\; \text{for } h \in \hat{\mathfrak{h}}, \tag{9.33b}$$

where m' denotes the number from $\{0, 1, \ldots, n-1\}$ congruent to m mod n.

Furthermore, denoting by $\mathfrak{n}_+$ (resp. $\mathfrak{n}_-$) the subalgebras of the strictly upper (resp. strictly lower) triangular matrices of sl_n, we have the triangular decomposition: $sl_n = \mathfrak{n}_- \oplus \mathfrak{h} \oplus \mathfrak{n}_+$. The corresponding *triangular decomposition* of $\hat{sl}_n$ is constructed as follows. Put $\hat{\mathfrak{n}}_+ = \mathfrak{n}_+ + \sum_{k>0} t^k \, sl_n$, $\hat{\mathfrak{n}}_- = \mathfrak{n}_- + \sum_{k>0} t^{-k} sl_n$. Then we have:

$$\hat{sl}_n = \hat{\mathfrak{n}}_+ \oplus \hat{\mathfrak{h}} \oplus \hat{\mathfrak{n}}_-.$$

Definition 9.5. For a given $\lambda \in \hat{\mathfrak{h}}^*$, called the *highest weight*, we define the *highest weight representation* π_λ of the Lie algebra $\hat{sl}_n$ as an irreducible representation on a vector space $L(\lambda)$ which admits a nonzero vector v_λ, called a *highest weight vector*, such that

$$\pi_\lambda(\hat{\mathfrak{n}}_+)v_\lambda = 0\,,$$

$$\pi_\lambda(h)v_\lambda = \lambda(h)v_\lambda \quad \text{for } h \in \hat{\mathfrak{h}}\,. \tag{9.34}$$

Remark 9.3. A general argument (as used in Lecture 3 for the Virasoro algebra) proves the existence and uniqueness of $L(\lambda)$ for all $\lambda \in \hat{\mathfrak{h}}^*$. ∎

We see that we have a representation of $\hat{s\ell}_n$ in $F^{(m)}$ which, with $v_\lambda = \psi_m$, satisfies the requirements (9.34) for a highest weight representation (here we use Proposition 9.1(b)). This representation is unitary due to Proposition 9.1(d) and the unitarity of the representation of $\mathfrak{a}_\infty$ in each $F^{(m)}$.

Recall that the representation $F^{(m)}$ is irreducible under $\hat{g\ell}'_n$ since the latter contains the $\Lambda_j (j \in \mathbb{Z})$, i.e., the generators of the oscillator algebra $\mathscr{A}$. To determine whether Λ_j belongs to the traceless subalgebra $\hat{s\ell}'_n$ of $\hat{g\ell}'_n$, we recall from Proposition 9.1(c) that Λ_j is the image in $\mathfrak{a}_\infty$ of the j-th power of the $n \times n$ matrix

$$a = \begin{bmatrix} 0 & 1 & 0 & \cdots\cdots & 0 \\ 0 & 0 & 1 & \cdots\cdots & 0 \\ \cdots\cdots\cdots\cdots\cdots \\ \cdots\cdots\cdots\cdots\cdots \\ 0 & 0 & 0 & \cdots\cdots & 1 \\ t & 0 & 0 & \cdots\cdots & 0 \end{bmatrix}.$$

It is easily checked that a^j is traceless when j is not an integral multiple of n, while $a^{sn} = t^s I$. Hence $\Lambda_j \in \hat{s\ell}'_n$ when j is not an integral multiple of n, while $\hat{s\ell}'_n$ commutes with all Λ_j for which j is an integral multiple of n. Since the subspace

$$F^{(m)}_{(0)} = \{v \in F^{(m)} \mid \hat{r}(\Lambda_{sn})v = 0,\ s = 1, 2, \ldots\} \tag{9.35}$$

is spanned by the vectors obtained by the repeated applications on ψ_m of the Λ_{-j} for $j \neq sn\,(s \in \mathbb{N})$, it is an irreducible invariant subspace for $\hat{s\ell}'_n$. It follows immediately that $F^{(m)}_{(0)}$ is invariant and irreducible also under $\hat{s\ell}_n$.

We can determine $F^{(m)}_{(0)}$ in more explicit form with the help of the boson-fermion equivalence under which $F^{(m)}$ is isomorphic to $B^{(m)} = \mathbb{C}[x_1, x_2, \ldots]$. The image $B^{(m)}_{(0)}$ of $F^{(m)}_{(0)}$ is easily determined since Λ_j is represented by $\partial/\partial x_j$ for $j > 0$. Clearly

$$B_{(0)}^{(m)} = \mathbb{C}[x_j \mid j \geqslant 1, \quad j \text{ is not a multiple of } n],$$

where on the right-hand side we have the subspace of polynomials in those variables x_j for which j is not a multiple of n.

Thus, $F_{(0)}^{(m)} = B_{(0)}^{(m)}$ is a unitary highest weight representation space of $\hat{s\ell}_n$ with highest weight $\omega_{m'}$, where $m' \in \{0, 1, \ldots, n-1\}$ and m' is congruent to m modulo n, i.e. $F_{(0)}^{(m)} \cong L(\omega_{m'})$. The representation $L(\omega_m)$ $(0 \leqslant m \leqslant n-1)$ is called the m-th *fundamental representation* of $\hat{s\ell}_n$.

We can clearly take tensor products of the fundamental representations and the highest component will have a highest weight equal to the sum of the individual highest weights. Summarizing, we have proved:

Proposition 9.3. The representations

$$L(k_0\omega_0 + k_1\omega_1 + \ldots + k_{n-1}\omega_{n-1})$$

of $\hat{s\ell}_n$, where $k_i \in \mathbb{Z}_+$, $0 \leqslant i \leqslant n-1$, are unitary. ∎

Remark 9.4. Proposition 9.3 is a special case of a theorem of Garland [1978] and Kac-Peterson [1984b] for non-twisted affine algebras and general Kac-Moody algebras respectively. ∎

Theorem 9.1. The representation $L(\lambda)$ of $\hat{s\ell}_n$ is unitary if and only if $\lambda(h_i) \in \mathbb{Z}_+$ for $i = 0, \ldots, n-1$ and $\lambda(d) \in \mathbb{R}$.

Proof. The 'if' part follows from Proposition 9.3 since we can make $\lambda(d) = 0$ by adding to d an arbitrary real constant. To see the 'only if' part, put

$$e_0 = e_{n,1}(1), \quad f_0 = e_{1,n}(-1),$$

$$e_i = e_{i,i+1}(0), \quad f_i = e_{i+1,i}(0) \quad \text{for } i = 1, \ldots, n-1.$$

Then $\{e_i, h_i, f_i\}$ form a standard basis of $s\ell_2$ for each i, and $\omega(e_i) = f_i$ (see (9.10)). But the only unitary irreducible representations of $s\ell_2$ with this involution are the finite-dimensional ones. Thus, unitarity of $L(\lambda)$ implies that $\lambda(h_i) \in \mathbb{Z}_+$ for all i, and, in particular, that $\lambda(M) \in \mathbb{Z}_+$. ∎

Since $M = h_0 + \ldots + h_{n-1}$, we deduce

Corollary 9.1. If the representation $L(\lambda)$ of $\hat{s\ell}_n$ is unitary, then $\lambda(M) \in \mathbb{Z}_+$. ∎

The integer $\lambda(M)$ is called the *level* of $L(\lambda)$.

LECTURE 10

10.1. Nonabelian generalization of Virasoro operators: the Sugawara construction

In Lecture 9 we constructed the affine algebra $\hat{g\ell}'_n$ starting from the finite dimensional Lie algebra $g\ell_n$. The restriction of the bilinear form (9.18) on $g\ell_n$ to its subalgebra $s\ell_n$ remains nondegenerate and we have the associated affine algebra $\hat{s\ell}'_n$. In fact for any finite-dimensional Lie algebra $\mathfrak{g}$ which has an invariant symmetric nondegenerate bilinear form $(\cdot \,|\, \cdot)$ there is a corresponding affine algebra $\hat{\mathfrak{g}}' = \tilde{\mathfrak{g}} \oplus \mathbb{C}M$, with commutation relations·

$$[x(k),\, M] = 0\,,$$
$$[x(k), y(m)] = [x, y](k + m) + k\delta_{k,-m}(x \,|\, y)M\,, \tag{10.1}$$

in analogy with (9.16). In the special case where $\mathfrak{g}$ is the one-dimensional abelian Lie algebra, (10.1) evidently reduces to the commutation relations (2.2) of the oscillator algebra $\mathscr{A}$. Thus we can view $\hat{\mathfrak{g}}'$ as a nonabelian generalization of $\mathscr{A}$.

We define $\hat{\mathfrak{g}} = \hat{\mathfrak{g}}' \oplus \mathbb{C}d$ in the same way as in Definition 9.4 for $\mathfrak{g} = s\ell_n$, and, as in §9.4, introduce the Cartan subalgebra $\hat{\mathfrak{h}} = \mathfrak{h} \oplus \mathbb{C}M \oplus \mathbb{C}d$, where $\mathfrak{h}$ is a Cartan subalgebra of $\mathfrak{g}$. Then $\hat{\mathfrak{g}}$ carries a nondegenerate symmetric invariant bilinear form $(\cdot \,|\, \cdot)$ defined by (9.26), with $\operatorname{tr} ab$ replaced by $(a \,|\, b)$.

An important example of $\mathfrak{g}$ which has an invariant symmetric nondegenerate bilinear form is a *reductive* Lie algebra (= a direct sum of simple Lie algebras and an abelian Lie algebra). Then one has a triangular decomposition of $\mathfrak{g}$, and in the same way as for $\hat{s\ell}_n$ in Section 9.4 one defines the associated triangular decomposition of $\hat{\mathfrak{g}}$ and irreducible highest weight representations $L(\lambda)$. It is easy to show that $L(\lambda)$ remains irreducible when restricted to $\hat{\mathfrak{g}}'$ (see e.g. Kac [1983], §9.10). In what follows, $\mathfrak{g}$ is assumed to be a (finite-dimensional) reductive Lie algebra.

We shall show in this lecture that we can use the generators of $\hat{\mathfrak{g}}'$ to construct representations of *Vir* in a nonabelian generalization of Virasoro's construction described in Lecture 2. Before doing this we shall need to collect some simple properties of $\mathfrak{g}$ and $\hat{\mathfrak{g}}'$.

Let $\{u_i \,|\, i = 1, \ldots, \dim \mathfrak{g}\}$ be a basis in $\mathfrak{g}$ and let $\{u^i \,|\, i = 1, \ldots, \dim \mathfrak{g}\}$ be the dual basis, so that

$$(u_i \,|\, u^j) = \delta_{ij} \,.$$

Then for any $x \in \mathfrak{g}$ we have

$$x = (x \,|\, u_i)u^i = (x \,|\, u^i)u_i \,. \tag{10.2}$$

Note that in (10.2) and subsequently we have adopted the convention that a repeated upper and lower index is summed over the dimension of $\mathfrak{g}$.

Let $\Omega_0 = u_i u^i$ be the *Casimir operator* of $\mathfrak{g}$. It is easily seen to be independent of the choice of dual bases $\{u_i\}$ and $\{u^i\}$. In particular, we have

$$[u_i, \, u^i] = 0 \,. \tag{10.3}$$

Lemma 10.1. $[\mathfrak{g}, \Omega_0] = 0.$

Proof.   Let $x \in \mathfrak{g}$. Then,

$$
\begin{aligned}
[x, \, u_i u^i] &= [x, u_i]u^i + u_j[x, u^j] \\
&= [x, u_i]u^i + u_j([x, u^j] \,|\, u_i)u^i && \text{(by (10.2))} \\
&= [x, u_i]u^i - u_j(u^j \,|\, [x, u_i])u^i && \text{(by invariance)} \\
&= [x, u_i]u^i - [x, u_i]u^i && \text{(by (10.2))} \\
&= 0 \,. && \blacksquare
\end{aligned}
$$

The Casimir operator Ω_0 acts as a multiplication by a scalar on any highest weight representation V of $\mathfrak{g}$ with highest weight λ (due to Lemma 10.1). Recall that one has:

$$\Omega_0 = (\lambda \,|\, \lambda + 2\overline{\rho})I \,, \tag{10.4}$$

where $\bar{\rho}$ is the sum of the fundamental weights of $\mathfrak{g}$. (Recall that (10.4) is proved by choosing a basis $\{u_i\}$ of $\mathfrak{g}$ consistent with the root space decomposition, replacing in Ω_0 expressions $e_\alpha e_{-\alpha}$ by $e_{-\alpha}e_\alpha + h_\alpha$ for $\alpha > 0$ and using the fact that $2\bar{\rho}$ is a sum of positive roots; then it is immediate that $\Omega_0(v_\lambda) = (\lambda \,|\, \lambda + 2\bar{\rho})v_\lambda$ if v_λ is a highest weight vector.)

Since

$$(u_i(m) \,|\, u^j(n)) = \delta_{ij}\delta_{m,-n}\,, \tag{10.5}$$

from (10.1) and (10.3) we have:

$$[u_i(m),\, u^i(n)] = m\, M\, \delta_{m,-n} \dim \mathfrak{g}\,. \tag{10.6}$$

Lemma 10.2. $[x, u_i](m)u^i(n) + u_i(m)[x, u^i](n) = 0$, for $x \in \mathfrak{g}$ and $m, n \in \mathbb{Z}$.

Proof is similar to that of Lemma 10.1. ∎

Lemma 10.3. Let the Lie algebra $\mathfrak{g}$ be simple or abelian. Then,

$$[u_i, [u^i, x]] = 2gx = [u^i, [u_i, x]]\,,$$

where g is a scalar (the factor 2 is inserted for convenience) and $g = 0$ if $\mathfrak{g}$ is abelian.

Proof. The lemma clearly holds if $\mathfrak{g}$ is abelian. Since Ω_0 commutes with $\mathfrak{g}$, it acts as a scalar in every irreducible representation of $\mathfrak{g}$, in particular in the adjoint representation if $\mathfrak{g}$ is simple. Denoting this scalar by $2g$ and noting that in the adjoint representation

$$\Omega_0 = (\operatorname{ad} u_i)(\operatorname{ad} u^i) = (\operatorname{ad} u^i)(\operatorname{ad} u_i)$$

completes the proof. ∎

Definition 10.1. A representation of the affine algebra $\hat{\mathfrak{g}}'$ on a vector space V is called *admissible* if for every $v \in V$ we have $x(k)(v) = 0$ for all $x \in \mathfrak{g}$ and all $k \gg 0$. It is clear that $L(\lambda)$ is an admissible representation of $\hat{\mathfrak{g}}'$.

Proposition 10.1. Let $\mathfrak{g}$ be a simple or abelian Lie algebra with a non-degenerate symmetric bilinear invariant form $(\cdot \,|\, \cdot)$, and let $\hat{\mathfrak{g}}'$ be the corresponding affine algebra with commutation relations (10.1). Consider an admissible representation of $\hat{\mathfrak{g}}'$ on a vector space V. Then the operators T_k defined by

$$T_k = \frac{1}{2} \sum_{j \in \mathbb{Z}} :u_i(-j)\,u^i(j+k): \tag{10.7}$$

for $k \in \mathbb{Z}$, satisfy the commutation relations

$$[x(n), T_k] = (M+g)n\,x(n+k). \tag{10.8}$$

Remark 10.1. The *normal ordering* :: in (10.7) means as usual that the order within the colons is to be preserved when $-j \leqslant j+k$ and reversed otherwise. Then the series in (10.7) will always be finite when applied to any $v \in V$ (for admissible V). Also, since $u_i(m)u^i(n)$ is independent of the choice of dual bases, this holds for the T_k as well. Finally, it follows from (10.6) that we may drop the sign of normal ordering in (10.7) if $k \neq 0$. ∎

Proof of Proposition 10.1.   We shall use the cutoff procedure described in Lecture 2. Let

$$T_k(\epsilon) = \frac{1}{2} \sum_{j \in \mathbb{Z}} :u_i(-j)u^i(j+k): \psi(\epsilon j),$$

where $\psi(x)$ is the cutoff function defined in (2.13a). Then

$$[x(n), T_k(\epsilon)] = \frac{1}{2} \sum_j [x, u_i](n-j)u^i(j+k)\psi(\epsilon j)$$

$$+ \frac{1}{2} \sum_j u_i(-j)[x, u^i](j+k+n)\psi(\epsilon j) + \frac{1}{2}nM\,x(n+k)\psi(\epsilon n)$$

$$+ \frac{1}{2}nM\,x(n+k)\psi(\epsilon(n+k)). \tag{10.9}$$

We split the first sum in (10.9) into terms for which $j \geqslant (n-k)/2$ (hence in normal order) and those for which $j < (n-k)/2$. The latter are replaced by normal-ordered terms with the help of the commutation relation

$$[[x, u_i](m), u^i(n)] = [[x, u_i], u^i](m+n), \tag{10.10a}$$

which follows immediately from (10.1), since the coefficient of M vanishes by the invariance of the bilinear form and (10.3). We now apply Lemma 10.3 to the right-hand side of (10.10a). The second sum in (10.9) is similarly split into normal-ordered terms satisfying $j \geqslant -(n+k)/2$ and the remainder for which $j < -(n + k)/2$. For the latter we use the dual of (10.10a):

$$[[x, u^i](m), u_i(n)] = [[x, u^i], u_i](m + n) \qquad (10.10b)$$

and Lemma 10.3 once more. In this way we get

$$[x(n), T_k(\epsilon)] = \frac{1}{2} \sum_j :[x, u_i](n - j)u^i(j + k): \psi(\epsilon j)$$

$$+ \frac{1}{2} \sum_j :u_i(-j)[x, u^i](j + k + n): \psi(\epsilon j) + \frac{1}{2}nM\, x(n + k)\psi(\epsilon n)$$

$$+ \frac{1}{2}nM\, x(n + k)\psi(\epsilon(n + k)) + gx(n + k) \sideset{}{'}\sum_j \psi(\epsilon j),$$

where $\sum'$ is taken over $-(n + k)/2 \leqslant j < (n - k)/2$. Making the transformation $j \rightarrow j + n$ in the first sum, we obtain in the limit $\epsilon \rightarrow 0$:

$$[x(n), T_k] = \frac{1}{2} \sum_j :[x, u_i](-j)u^i(j + k + n):$$

$$+ \frac{1}{2} \sum_j :u_i(-j)[x, u^i](j + k + n): + n(g + M)x(n + k).$$

By Lemma 10.2 the combined two sums vanish, which proves the proposition. ■

Theorem 10.1. Under the same hypotheses as in Proposition 10.1 we have

$$[T_n, T_k] = (M + g)(n - k)T_{n+k} + \delta_{n,-k}\frac{(n^3 - n)}{12}\,(\dim \mathfrak{g})M(M + g)\,. \quad (10.11)$$

Proof.

$$[T_n(\epsilon), T_k] = \frac{1}{2} \sum_{j \in \mathbb{Z}} [u_i(-j)u^i(j + n), T_k]\psi(\epsilon j)$$

$$= \frac{1}{2}(M + g) \sum_j (-j)u_i(k - j)u^i(j + n)\psi(\epsilon j)$$

$$+ \frac{1}{2}(M + g) \sum_j (j + n)u_i(-j)u^i(j + k + n)\psi(\epsilon j)$$

by Proposition 10.1. We now reorder terms with the help of (10.6) to get normal-ordered expressions:

$$[T_n(\epsilon), T_k] = \frac{1}{2}(M+g)\sum_j (-j) : u_i(k-j)u^i(j+n) : \psi(\epsilon j)$$

$$+ \frac{1}{2}(M+g)\sum_j (j+n) : u_i(-j)u^i(j+k+n) : \psi(\epsilon j)$$

$$+ \frac{1}{2}M(M+g)(\dim \mathfrak{g})\delta_{n,-k}\left\{ -\sum_{j=-1}^{-n} j(n+j)\psi(\epsilon j)\right\}.$$

Making the transformation $j \to j+k$ in the first sum, we obtain (10.11) on taking the limit $\epsilon \to 0$. ■

Remark 10.2. Let $\mathfrak{g}$ be simple. We normalize the bilinear form $(\cdot \,|\, \cdot)$ by choosing long roots to have square length 2. (Note that for $\mathfrak{g} = sl_n$, $(x \,|\, y) = \operatorname{tr} xy$.) With this normalization, it can be shown in the same way as for $\hat{sl}_n$, that if $L(\Lambda)$ is a unitary highest weight representation of $\hat{\mathfrak{g}}$ then the central element M is represented by mI, where m is a non-negative integer. In this normalization of $(\cdot \,|\, \cdot)$ the number g, which is half the eigenvalue of Ω_0 in the adjoint representation, is a positive integer. It is known as the *dual Coxeter number*. Using (10.4) for $\lambda = \theta$, the highest root, we find:

$$g = 1 + (\theta \,|\, \overline{\rho}).$$

The concrete values of g are given below. ■

Table 10.1

Simple Lie algebra $\mathfrak{g}$	dim $\mathfrak{g}$	Dual Coxeter number g
A_ℓ	$\ell^2 + 2\ell$	$\ell + 1$
B_ℓ	$2\ell^2 + \ell$	$2\ell - 1$
C_ℓ	$2\ell^2 + \ell$	$\ell + 1$
D_ℓ	$2\ell^2 - \ell$	$2\ell - 2$
E_6	78	12
E_7	133	18
E_8	248	30
F_4	52	9
G_2	14	4

Proposition 10.1 and Theorem 10.1 now give us:

Corollary 10.1. Let V be an admissible representation of $\hat{\mathfrak{g}}'$ ($\mathfrak{g}$ is simple or abelian) such that $M = mI$ with $m \neq -g$. Then:
(a) We can define

$$L_k = \frac{1}{(m+g)} T_k \qquad (10.12)$$

so that we have:

$$[L_k, L_n] = (k-n)L_{k+n} + \delta_{k+n,0} \frac{k^3 - k}{12} \frac{m \dim \mathfrak{g}}{m+g}, \qquad (10.13a)$$

$$[L_k, x(n)] = -nx(n+k). \qquad (10.13b)$$

In particular, putting $d = -L_0$, V can be extended to a representation of $\hat{\mathfrak{g}}$.
(b) If V is unitary, we thus obtain a unitary representation of Vir in V with central charge

$$c = \frac{m \dim \mathfrak{g}}{m+g}. \qquad (10.14)$$

Proof. The only thing that remains to check is unitarity. For that choose a basis $\{v_j\}$ of the compact form of $\mathfrak{g}$ (= the fixed point set of $-\omega$, where ω is the compact antilinear anti-involution of $\mathfrak{g}$) such that $(v_j \mid v_k) = -\delta_{jk}$, and put $u_j = u^j = i\,v_j$. Then:

$$\omega(L_n) = \omega\left(\sum_i \sum_{j \in \mathbb{Z}} u_i(-j)u_i(j+n)\right) = \sum_i \sum_{j \in \mathbb{Z}} u_i(-j-n)u_i(j) = L_{-n}$$

if $n \neq 0$ (see Remark 10.1). A similar calculation applies in the case $n = 0$. ∎

Remark 10.3. Note from Table 10.1 that $c \geqslant 1$ in (10.14).

Let

$$\mathfrak{g} = \bigoplus_{i=0}^{\ell} \mathfrak{g}_i \qquad (10.15)$$

be a reductive Lie algebra with center $\mathfrak{g}_0$. We can choose bases $\{u_j\}, \{u^j\}$ in $\mathfrak{g}$ which respect the decomposition (10.15) and are dual with respect to an invariant, symmetric bilinear form $(\cdot \mid \cdot)$. We assume that this form is properly normalized when restricted to each component of $\mathfrak{g}$ (Remark 10.2). Note that $\hat{\mathfrak{g}}'$ is the direct sum of the $\hat{\mathfrak{g}}'_i$. Corresponding to the unitary highest weight representation of $\hat{\mathfrak{g}}'_i$ in $L(\Lambda_i)$ we have a family of operators $L_k^{(i)}$ ($k \in \mathbb{Z}$) which satisfy the Virasoro algebra, where m_i is the level of $L(\Lambda_i)$ and g_i the dual Coxeter number of $\mathfrak{g}_i$. Furthermore, $\hat{\mathfrak{g}}'$ acts on the tensor product of the $L(\Lambda_i)$ in the usual way, so that the $L_k^{(i)}$ commute for different i. Hence, defining for each $k \in \mathbb{Z}$

$$L_k = \sum_{i=0}^{\ell} L_k^{(i)}, \tag{10.16}$$

we get:

Corollary 10.2. The L_k ($k \in \mathbb{Z}$) form a unitary representation of the Virasoro algebra with central charge given by

$$c = \sum_{i=0}^{\ell} \frac{(\dim \mathfrak{g}_i)m_i}{m_i + g_i}. \tag{10.17}$$

■

Remark 10.4. The construction of Theorem 10.1 is a discrete counterpart of the Sugawara [1968] construction. The earliest reference that we know in which the central charge is given correctly (in the case of $su(n)$) is the paper Dashen-Frishman [1975]. In the following we shall follow standard practice and refer to this as the Sugawara construction.   ■

Note that

$$T_0 = \frac{1}{2}u_i u^i + \sum_{j>0} u_i(-j)u^i(j). \tag{10.18}$$

The operator T_0 is closely related to the Casimir operator Ω of $\hat{\mathfrak{g}}$ introduced by Kac [1974] in the framework of general Kac-Moody algebras:

Proposition 10.2. Let $\mathfrak{g}$ be a simple finite-dimensional Lie algebra.

(a) The operator Ω defined by

$$\Omega = 2(M + g)d + 2T_0 \tag{10.19}$$

commutes with every element of $\hat{\mathfrak{g}}$.

(b) On $L(\Lambda)$ the eigenvalue of Ω is $(\Lambda + 2\rho \,|\, \Lambda)$ where $\rho = \sum_i \omega_i$ is the sum of the fundamental weights (which are defined in the same way as for $\hat{s\ell}_n$ in §9.4).

Proof. (a) We see that $[d, \Omega] = 0$ by using the representation (10.18) for T_0 and (9.25). The equality $[x(n), \Omega] = 0$ follows from the application of (9.25) and (10.8).

(b) Applying Ω to the highest weight vector we see from (10.18) that the only nonzero contribution is from $2(M + g)d + \Omega_0$. Using (10.4), we get the formula immediately since $\rho = \overline{\rho} + g\omega_0$. ∎

10.2.   The Goddard-Kent-Olive construction

In the previous section (see Remark 10.3 and Corollary 10.2) we saw that the unitary representation of *Vir* obtained from $\hat{g}'$, where g is a reductive Lie algebra, has a central charge which is always greater than or equal to 1. Goddard-Kent-Olive [1985] have found a way to construct unitary representations of *Vir* for which the central charge is less than 1.

We take a reductive Lie algebra g and a reductive subalgebra p of g. Given a unitary highest weight representation of g we can use the Sugawara construction to form a unitary representation of *Vir*, *viz.* $\{L_k^g \,|\, k \in \mathbb{Z}\}$ with central charge c_g from the generators of $\hat{g}'$. We can construct a second unitary representations of *Vir* $\{L_k^p \,|\, k \in \mathbb{Z}\}$ with central charge c_p from the generators of $\hat{p}'$. (We have to be careful to follow the normalization of Remark 10.2 separately for $\hat{g}'$ and $\hat{p}'$.)

Theorem 10.2. The operators

$$L_k = L_k^g - L_k^p \quad (k \in \mathbb{Z}) \tag{10.20}$$

form a unitary representation of the Virasoro algebra with central charge

$$c = c_g - c_p . \tag{10.21}$$

Proof. We first show that L_k commutes with L_n^p. This follows from the fact that L_k commutes with $x(n)$ when $x(n) \in \hat{p}'$:

$$[L_k, x(n)] = [L_k^{\mathfrak{g}}, x(n)] - [L_k^{\mathfrak{p}}, x(n)]$$
$$= nx(n+k) - nx(n+k) = 0.$$

Thus

$$[L_k, \hat{\mathfrak{p}}'] = 0 \quad (k \in \mathbb{Z}), \qquad (10.22)$$

and hence

$$[L_k, L_n^{\mathfrak{p}}] = 0 \quad (k, n \in \mathbb{Z}).$$

It follows now that

$$[L_k, L_n] = [L_k, L_n^{\mathfrak{g}}] = [L_k^{\mathfrak{g}}, L_n^{\mathfrak{g}}] - [L_k^{\mathfrak{p}}, L_n^{\mathfrak{g}}]$$
$$= [L_k^{\mathfrak{g}}, L_n^{\mathfrak{g}}] - [L_k^{\mathfrak{p}}, L_n + L_n^{\mathfrak{p}}]$$
$$= [L_k^{\mathfrak{g}}, L_n^{\mathfrak{g}}] - [L_k^{\mathfrak{p}}, L_n^{\mathfrak{p}}]$$
$$= (k-n)L_{k+n} + \delta_{k+n,0} \frac{(k^3 - k)}{12}(c_{\mathfrak{g}} - c_{\mathfrak{p}}).$$

The unitarity of this representation of *Vir* follows from Corollary 10.1. ∎

We now consider the special case where $\mathfrak{p}$ is a simple Lie algebra and $\mathfrak{g} = \mathfrak{p} \oplus \mathfrak{p}$. We consider two representations of $\hat{\mathfrak{p}}$ on $L(\Lambda)$ and $L(\Lambda')$ with levels m and m' respectively. Then the action of $\hat{\mathfrak{g}}' = \hat{\mathfrak{p}} \oplus \hat{\mathfrak{p}}$ on $L(\Lambda) \otimes L(\Lambda')$ is given by:

$$(x(n) \oplus y(m))(v \otimes w) = (x(n)(v)) \otimes w + v \otimes (y(m)(w))$$

for $v \otimes w \in L(\Lambda) \otimes L(\Lambda')$. From this it is clear that the Sugawara construction on $\hat{\mathfrak{g}}'$ gives

$$L_k^{\mathfrak{g}} = L_k^{\mathfrak{p}} \otimes 1 + 1 \otimes L_k^{\mathfrak{p}},$$

with central charge $c_{\mathfrak{g}} = (\dim \mathfrak{p})(m/(m+g) + m'/(m'+g))$. Also $\hat{\mathfrak{p}}'$ has the "diagonal" action on $L(\Lambda) \otimes L(\Lambda')$ given by

$$x(n)(v \otimes w) = (x(n)(v)) \otimes w + v \otimes (x(n)(w)).$$

(This is equivalent to embedding $\hat{\mathfrak{p}}'$ diagonally in $\hat{\mathfrak{p}} \oplus \hat{\mathfrak{p}}'$, i.e. $x(n) \to x(n) \oplus x(n)$.) The level of this representation of $\hat{\mathfrak{p}}'$ is clearly $m + m'$. This gives rise to Virasoro operators $L_k^{\mathfrak{p}}$ with central charge $= (\dim \mathfrak{p})(m + m')/(m + m' + g)$.

Hence, $L_k = L_k^{\mathfrak{g}} - L_k^{\mathfrak{p}}$ gives, by Theorem 10.2, a representation of *Vir* with central charge

$$c = (\dim \mathfrak{p}) \left(\frac{m}{m+g} + \frac{m'}{m'+g} - \frac{m+m'}{m+m'+g} \right).$$

From (10.19) we observe that

$$
L_0 = \left(\frac{1}{2(m+g)} \Omega^{\mathfrak{p}} \Big|_{L(\Lambda)} - d \right) \otimes 1 + 1 \otimes \left(\frac{1}{2(m'+g)} \Omega^{\mathfrak{p}} \Big|_{L(\Lambda')} - d \right)
$$
$$
- \left(\frac{1}{2(m+m'+g)} \Omega^{\mathfrak{p}} \Big|_{L(\Lambda) \otimes L(\Lambda')} - d \otimes 1 - 1 \otimes d \right).
$$

So by Proposition 10.2(b) we have

$$
L_0 = \frac{1}{2} \left(\frac{(\Lambda \,|\, \Lambda + 2\rho)}{m+g} + \frac{(\Lambda' \,|\, \Lambda' + 2\rho)}{m'+g} - \frac{\Omega^{\mathfrak{p}}}{m+m'+g} \right).
$$

We summarize these results in the following proposition.

Proposition 10.3. Let $\mathfrak{p}$ be a simple Lie algebra with dual Coxeter number g. Let $\{u_i\}$ and $\{u^i\}$ be dual bases of $\mathfrak{p}$, and consider two highest weight unitary representations $L(\Lambda), L(\Lambda')$ of $\hat{\mathfrak{p}}$ with levels m and m'. Then

(a) The following operators ($k \in \mathbb{Z}$)

$$
L_k = \left(\frac{1}{2(m+g)} - \frac{1}{2(m+m'+g)} \right) \sum_{j \in \mathbb{Z}} :u_i(-j)u^i(j+k): \otimes 1
$$
$$
+ \left(\frac{1}{2(m'+g)} - \frac{1}{2(m+m'+g)} \right) \sum_{j \in \mathbb{Z}} 1 \otimes :u_i(-j)u^i(j+k):
$$
$$
- \frac{1}{(m+m'+g)} \sum_{j \in \mathbb{Z}} u_i(-j) \otimes u^i(j+k), \tag{10.23}
$$

on the space $L(\Lambda) \otimes L(\Lambda')$ form a unitary representation of the Virasoro algebra with central charge

$$c = (\dim \mathfrak{p}) \left(\frac{m}{m+g} + \frac{m'}{m'+g} - \frac{m+m'}{m+m'+g} \right). \tag{10.24}$$

(b) $L_0 = \dfrac{1}{2} \left(\dfrac{(\Lambda \,|\, \Lambda + 2\rho)}{m + g} + \dfrac{(\Lambda' \,|\, \Lambda' + 2\rho)}{m' + g} - \dfrac{\Omega}{m + m' + g} \right),$ (10.25)

where Ω is the Casimir of $\hat{\mathfrak{p}}$.

(c) $[L_k, \hat{\mathfrak{p}}'] = 0 \,,$ (10.26)

i.e. the L_k are intertwining operators for the representation of $\hat{\mathfrak{p}}'$ on $L(\Lambda) \otimes L(\Lambda')$. ∎

LECTURE 11

11.1. $\hat{s\ell}_2$ and its Weyl group

In this lecture we shall discuss the character formula for a highest weight representation of the simplest affine algebra $\hat{s\ell}_2$. The rather technical results described in this lecture are a necessary preliminary to Lecture 12, where we shall, among other things, give the proof of Lemma 8.6 and thereby complete the proof of the Kac determinant formula.

In the notation of Lecture 9 the Cartan subalgebra $\hat{\mathfrak{h}}$ of $\hat{s\ell}_2$ is spanned by h_0, h_1 and d where

$$h_0 = M - e_{11}(0) + e_{22}(0),$$

$$h_1 = e_{11}(0) - e_{22}(0) = \begin{pmatrix} 1 & 0 \\ 0 & -1 \end{pmatrix}.$$

In the following we shall denote the matrix h_1 by α and choose as basis of $\hat{\mathfrak{h}}$ the elements $\alpha, M = h_0 + h_1$ and d:

$$\hat{\mathfrak{h}} = \mathbb{C}\alpha \oplus \mathbb{C}M \oplus \mathbb{C}d.$$

The bilinear form $(\cdot \mid \cdot)$ on $\hat{s\ell}_2$ is nondegenerate when restricted to $\hat{\mathfrak{h}}$; we see from (9.26) that

$$(\alpha \mid \alpha) = 2; \quad (M \mid d) = 1; \quad \text{all other pairs vanish.} \tag{11.1}$$

We shall identify $\hat{\mathfrak{h}}$ with $\hat{\mathfrak{h}}^*$ via this form.

In the highest weight representation $L(\Lambda)$ of $\hat{s\ell}_2$ (defined in Lecture 9) the action of $\hat{\mathfrak{h}}$ on the highest weight vector v_λ is given by

$$h(v_\lambda) = \lambda(h)v_\lambda = (\lambda \mid h)v_\lambda, \quad h \in \hat{\mathfrak{h}}.$$

113

The fundamental weights, defined by

$$\omega_i(h_j) = \delta_{ij}, \quad \omega_i(d) = 0 \quad (i, j = 0, 1),$$

are

$$\omega_0 = d, \quad \omega_1 = d + \frac{1}{2}\alpha. \tag{11.2}$$

As before, we shall denote by ρ the sum of the fundamental weights:

$$\rho \equiv \omega_0 + \omega_1 = 2d + \frac{1}{2}\alpha. \tag{11.3}$$

From (11.2) and Theorem 9.1 we see that $L(\lambda)$ is unitary if and only if λ is of the form

$$\lambda = md + \frac{1}{2}n\alpha + rM, \quad \text{where } r \in \mathbb{R}; \quad m, n \in \mathbb{Z}_+; \quad m \geqslant n. \tag{11.4}$$

Then

$$M(v_\lambda) = (\lambda \mid M)v_\lambda = mv_\lambda, \tag{11.5}$$

and so $L(\lambda)$ is a level m representation.

One of the ingredients which we shall require for the character formula is the Weyl group of $\hat{s\ell}_2$. The Weyl group W of a Lie algebra $\mathfrak{g}$ is the group of those automorphisms of a Cartan subalgebra of $\mathfrak{g}$ which are restrictions of conjugations by elements of G, the Lie group corresponding to $\mathfrak{g}$. In our case this means that the Weyl group is the quotient of the subgroup of $SL_2(\mathbb{C}[t, t^{-1}])$ which leaves $\hat{\mathfrak{h}}$ invariant under the adjoint action (defined in Lecture 9) by the subgroup which leaves $\hat{\mathfrak{h}}$ pointwise fixed. (For an equivalent definition, one in terms of reflections of the root system, see Kac [1983].)

The adjoint action of $SL_2(\mathbb{C}[t, t^{-1}])$ on α, M and d can be determined from (9.28). It is then easily checked that the Weyl group of $\hat{s\ell}_2$, which we shall denote by $\hat{W}$, is generated by two elements: one is conjugation by

$$\begin{bmatrix} 0 & 1 \\ -1 & 0 \end{bmatrix}$$

which we denote by r_α, and the second is conjugation by

$$\begin{bmatrix} t & 0 \\ 0 & t^{-1} \end{bmatrix}.$$

We shall denote conjugation by the k-th power of the latter matrix by $t_k(k \in \mathbb{Z})$. Then

$$\hat{W} = \{t_k, t_k r_\alpha \mid k \in \mathbb{Z}\}, \qquad (11.6)$$

since $r_\alpha^2 = 1$ and $t_k r_\alpha = r_\alpha t_{-k}$. The action of the elements of $\hat{W}$ on $\hat{\mathfrak{h}}$ is easily computed from (9.28):

$$\left.\begin{array}{lll} r_\alpha(\alpha) = -\alpha, & r_\alpha(M) = M, & r_\alpha(d) = d; \\[2mm] t_k(\alpha) = \alpha + 2kM, & t_k(M) = M, & t_k(d) = d - k\alpha - k^2 M \end{array}\right\}. \qquad (11.7)$$

The relations (11.7) give a matrix representation of r_α and t_k in $\hat{\mathfrak{h}}$. We shall denote by $\epsilon(w)$ the determinant of $w \in \hat{W}$ in this representation. We find that:

$$\epsilon(r_\alpha) = -1, \quad \epsilon(t_k) = 1, \quad \epsilon(t_k r_\alpha) = -1. \qquad (11.8)$$

11.2. The Weyl-Kac character formula and Jacobi-Riemann theta functions

Definition 11.1. We define the *character* of the representation $L(\lambda)$ to be the function

$$\mathrm{ch}_\lambda(h) = \mathrm{tr}_{L(\lambda)} \exp(h) \text{ for } h \in \hat{\mathfrak{h}}.$$

The study of the representations $L(\lambda)$ was started by Kac [1974] with the computation of $\mathrm{ch}_\lambda(h)$ in the framework of arbitrary Kac-Moody algebras. Namely, according to the so-called Weyl-Kac character formula,

$$\mathrm{tr}_{L(\lambda)} \exp(h) = \frac{\sum \epsilon(w) \exp(w(\lambda + \rho) \mid h)}{\sum \epsilon(w) \exp(w(\rho) \mid h)}, \qquad (11.9)$$

where the summations run over w in the Weyl group (whenever this series converges). For a proof of (11.9) the reader is referred to Chapter 10 of the book Kac [1983]. We shall now rewrite formula (11.9) in a more explicit form for the case of $\hat{s\ell}_2$ (arbitrary affine algebras can be treated similarly: see Chapter 12 of Kac [1983]).

The calculation of the denominator and numerator in (11.9) being similar, take

$$\mu = md + \frac{n}{2}\alpha + rM. \tag{11.10}$$

Recalling (11.6) and (11.8), we have:

$$\sum_{w \in \hat{W}} \epsilon(w) \exp(w(\mu) \,|\, h) = \sum_{k \in \mathbb{Z}} \exp(t_k(\mu) \,|\, h) - \sum_{k \in \mathbb{Z}} \exp(t_k r_\alpha(\mu) \,|\, h).$$

From (11.7) and (11.10), for each $k \in \mathbb{Z}$ we have:

$$t_k(\mu) = md + \left(\frac{1}{2}n - mk\right)\alpha + (r + kn - k^2 m)M, \tag{11.11a}$$

$$t_k r_\alpha(\mu) = md - \left(\frac{1}{2}n + mk\right)\alpha + (r - kn - k^2 m)M. \tag{11.11b}$$

We choose h to have the general form

$$h = 2\pi i \left(\frac{1}{2}z\alpha - \tau d + uM\right), \quad \text{where } \tau, z, u \in \mathbb{C}. \tag{11.12}$$

Then from (11.11a, b) and (11.1), for each $k \in \mathbb{Z}$ we have:

$$(t_k(\mu) \,|\, h) = 2\pi i \left[\left(\frac{1}{2}n - mk\right)z - (r + kn - k^2 m)\tau + mu\right], \tag{11.13a}$$

$$(t_k r_\alpha(\mu) \,|\, h) = 2\pi i \left[-\left(\frac{1}{2}n + mk\right)z - (r - kn - k^2 m)\tau + mu\right]. \tag{11.13b}$$

Replace k in (11.13a) by $n/2m - k$ (so that now $k \in n/2m + \mathbb{Z}$) and in (11.13b) by $-n/2m - k$ (so that $k \in -n/2m + \mathbb{Z}$); the right-hand sides of (11.13a, b) become respectively:

$$2\pi i \left(mkz + \left(mk^2 - \left(r + \frac{n^2}{4m}\right)\right)\tau + mu\right) \quad \left(k \in \frac{n}{2m} + \mathbb{Z}\right), \tag{11.14a}$$

$$2\pi i \left(mkz + \left(mk^2 - \left(r + \frac{n^2}{4m}\right)\right)\tau + mu\right) \quad \left(k \in -\frac{n}{2m} + \mathbb{Z}\right). \tag{11.14b}$$

Definition 11.2. We define $\Theta_{n,m}(\tau, z, u)$ by

$$\Theta_{n,m}(\tau, z, u) = \exp(2\pi i m u) \sum_{k \in \frac{n}{2m} + \mathbb{Z}} \exp(2\pi i m(k^2 \tau + kz)). \tag{11.15}$$

Remark 11.1. These are the celebrated Jacobi-Riemann theta functions. Clearly $\Theta_{n,m}(\tau, z, u)$ converges absolutely for τ in the upper half complex plane and arbitrary complex z and u.

We thus obtain

$$\sum_{w \in \hat{W}} \epsilon(w) \exp(w(\mu) \mid h) = q^{-(r + \frac{n^2}{4m})} (\Theta_{n,m}(\tau, z, u) - \Theta_{-n,m}(\tau, z, u)).$$

$$(11.16)$$

Here and further, q stands for $\exp 2\pi i \tau$.

In the numerator of (11.9) we have $\mu = \lambda + \rho = (m+2)d + \frac{1}{2}(n+1)\alpha + rM$, while in the denominator we have $\mu = \rho = 2d + \frac{1}{2}\alpha$. Substituting in (11.6) we arrive at:

Proposition 11.1. For $\lambda = md + \frac{1}{2}n\alpha + rM$ as in (11.4) and $z, \tau, u \in \mathbb{C}$, Im $\tau > 0$, one has:

$$\mathrm{ch}_\lambda(h) \equiv \mathrm{tr}_{L(\lambda)} \exp\left[2\pi i \left(\frac{1}{2}z\alpha - \tau d + uM\right)\right]$$

$$= q^{-s_\lambda} \frac{\Theta_{n+1,m+2}(\tau, z, u) - \Theta_{-n-1,m+2}(\tau, z, u)}{\Theta_{1,2}(\tau, z, u) - \Theta_{-1,2}(\tau, z, u)}, \quad (11.17a)$$

where

$$s_\lambda = \frac{(n+1)^2}{4(m+2)} - \frac{1}{8} + r. \quad (11.17b)$$

■

Remark 11.2. Dividing the numerator and the denominator of (11.17a) by $1 - e^{-2\pi i z}$, one derives from (11.7a), as $u = 0$ and $z \to 0$, a formula for the q-dimension (= partition function):

$$\mathrm{tr}_{L(md+\frac{1}{2}n\alpha)} q^{-d} = \varphi(q)^{-3} \sum_{j \in \mathbb{Z}} (2j(m+2) + n + 1) q^{(m+2)j^2 + (n+1)j}.$$

■

The theta functions $\Theta_{n,m}(\tau, z, u)$ have an important multiplication rule which we shall require (Kac-Peterson [1984a]):

Proposition 11.2.

$$\Theta_{n,m}(\tau, z, u) \Theta_{n',m'}(\tau, z, u)$$

$$= \sum_j d_j^{(m,m',n,n')}(q) \Theta_{n+n'+2mj, m+m'}(\tau, z, u) \quad (j \in \mathbb{Z} \bmod(m+m')\mathbb{Z}),$$

where

$$d_j^{(m,m',n,n')}(q) = \sum_k q^{mm'(m+m')k^2} \quad \left(k \in \mathbb{Z} + \frac{m'n - mn' + 2jmm'}{2mm'(m+m')}\right) .$$

Proof.   Without loss of generality we can put $u = 0$. Then

$$\Theta_{n,m} = \sum_k q^{mk^2} \exp(2\pi imkz) \quad \left(k \in \frac{n}{2m} + \mathbb{Z}\right) ,$$

$$\Theta_{n',m'} = \sum_{k'} q^{m'k'^2} \exp(2\pi im'k'z) \quad \left(k' \in \frac{n'}{2m'} + \mathbb{Z}\right) ,$$

and

$$\Theta_{n,m}\Theta_{n',m'} = \sum_{k,k'} q^{mk^2 + m'k'^2} \exp[2\pi i(mk + m'k')z] .$$

Let $k = (n/2m) + i, k' = (n'/2m') + i'$, where $i, i' \in \mathbb{Z}$. Define s, s' by

$$(m+m')s = k - k' = \frac{nm' - n'm}{2mm'} + i - i' ,$$

$$(m+m')s' = mk + m'k' = (m+m')(k' + ms) .$$

Let $i - i' = (m+m')\ell + j$ where $\ell \in \mathbb{Z}$ and $j \in \mathbb{Z} \bmod(m+m')\mathbb{Z}$. Then

$$s \in \frac{nm' - n'm + 2mm'j}{2mm'(m+m')} + \mathbb{Z} , \quad s' \in \frac{n + n' + 2mj}{2(m+m')} + \mathbb{Z} .$$

This gives us a bijection between pairs (k, k') and triples (s, s', j). Noting that

$$mk^2 + m'k'^2 = mm'(m+m')s^2 + (m+m')s'^2 ,$$

we get:

$$\Theta_{n,m}\Theta_{n',m'} = \sum_j \left(\sum_s q^{mm'(m+m')s^2}\right)\left(\sum_{s'} q^{(m+m')s'^2} \exp 2\pi i(m+m')s'z\right)$$

which proves the proposition.   ∎

We shall now consider the character formula (11.17a, b) for the simplest nontrivial case of $\lambda = d$ ($m = 1, n = 0, r = 0$). We call this the *basic representation.* Since $s_d = -1/24$, we get

$$\mathrm{ch}_d(h) = q^{1/24} \frac{\Theta_{1,3} - \Theta_{-1,3}}{\Theta_{1,2} - \Theta_{-1,2}}.$$

We shall now use Proposition 11.2 to prove a much simpler formula for $\mathrm{ch}_d(h)$:

Proposition 11.3.

$$\mathrm{ch}_d(h) = \frac{\Theta_{0,1}}{\varphi(q)}, \tag{11.18}$$

where $\varphi(q)$ is the function defined in (2.8b).

Proof. We have to prove that

$$\Theta_{0,1}(\Theta_{1,2} - \Theta_{-1,2}) = q^{1/24}\,\varphi(q)(\Theta_{1,3} - \Theta_{-1,3}).$$

Applying Proposition 11.2 to the left-hand side we obtain

$$(\Theta_{1,3} - \Theta_{-1,3}) \left(\sum_{k \in -\frac{1}{12}+\mathbb{Z}} q^{6k^2} - \sum_{k' \in \frac{5}{12}+\mathbb{Z}} q^{6k'^2} \right)$$

$$= (\Theta_{1,3} - \Theta_{-1,3})q^{1/24} \sum_{j \in \mathbb{Z}} (-1)^j q^{(3j^2+j)/2}.$$

A famous identity due to Euler tells us that

$$\varphi(q) = \sum_{k \in \mathbb{Z}} (-1)^k q^{(3k^2+k)/2} \tag{11.19}$$

which completes the proof. ∎

Remark 11.3. The Weyl-Kac character formula (11.9) is often written in a different form:

$$\mathrm{tr}_{L(\lambda)} \exp(h) = \frac{\displaystyle\sum_{w} \epsilon(w) \exp[(w(\lambda + \rho) \,|\, h) - (\rho \,|\, h)]}{\displaystyle\prod_{\substack{\gamma \in \Delta \\ (\rho \,|\, \gamma) > 0}} (1 - \exp[-(\gamma \,|\, h)])^{\mathrm{mult}\,\gamma}}. \tag{11.20}$$

Here Δ is the set of roots and mult γ is the multiplicity of $\gamma \in \Delta$. (Recall that $\gamma \in \hat{\mathfrak{h}}$ is called a root if the equation $[h, x] = (\gamma \,|\, h)x$ has a nonzero solution for all $h \in \hat{\mathfrak{h}}$; the dimension of the space of solutions is called the multiplicity of γ.) Putting $\lambda = 0$ in (11.20) we obtain the so-called Weyl-Macdonald-Kac formula:

$$\prod_{\substack{\gamma \in \Delta \\ (\rho \,|\, \gamma) > 0}} (1 - \exp[-(\gamma \,|\, h)])^{\text{mult}\,\gamma} = \sum_w \epsilon(w) \exp[(w(\rho) \,|\, h) - (\rho \,|\, h)]\,. \quad (11.21)$$

Plugging (11.21) into (11.20) gives (11.9). In the case of $\hat{s\ell}_2$, we have $\Delta = \{\pm\alpha + kd, kd \,|\, k \in \mathbb{Z}\}$ and the multiplicities of all roots are 1. Putting $u = \exp[-(h_0 \,|\, h)]$, $v = \exp[-(h_1 \,|\, h)]$, it is easy to see that (11.21) turns into the classical Jacobi triple product identity:

$$\prod_{k \geqslant 1}(1 - u^{k-1}v^k)(1 - u^k v^{k-1})(1 - u^k v^k) = \sum_{j \in \mathbb{Z}}(-1)^j u^{\frac{j(j+1)}{2}} v^{\frac{j(j-1)}{2}}\,.$$

A self-contained proof of this identity will be given in §16.5. Putting $u = q, v = q^2$ in this identity gives the Euler identity (11.19). See Chapter 12 of Kac [1983] for more details. ∎

11.3. A character identity

The following character identity will play a crucial role in Lecture 12.

Proposition 11.4. For $\lambda = md + \frac{1}{2}n\alpha$ $(m \geqslant n \geqslant 0)$, one has

$$\text{ch}_d \,\text{ch}_\lambda = \sum_{k \in I} \psi_{m,n;k} \,\text{ch}_{d+\lambda-k\alpha}\,, \qquad (11.22a)$$

where

$$I = \left\{ k \in \mathbb{Z} \,\middle|\, -\frac{1}{2}(m+1-n) \leqslant k \leqslant \frac{1}{2}n \right\}, \qquad (11.22b)$$

and

$$\psi_{m,n;k} = (f_k^{(m,n)} - f_{n+1-k}^{(m,n)})/\varphi(q)\,, \qquad (11.22c)$$

where

$$f_k^{(m,n)} = \sum_{j \in \mathbb{Z}} q^{(m+2)(m+3)j^2 + ((n+1)+2k(m+2))j + k^2}\,. \qquad (11.22d)$$

Proof. From (11.17) and (11.18),

$$\varphi(q)\,\mathrm{ch}_d\,\mathrm{ch}_\lambda = \frac{\Theta_{0,1}\Theta_{n+1,m+2} - \Theta_{0,1}\Theta_{-n-1,m+2}}{\Theta_{1,2} - \Theta_{-1,2}} q^{-\frac{(n+1)^2}{4(m+2)}+\frac{1}{8}}.$$

By Proposition 11.2:

$$\Theta_{0,1}\Theta_{n+1,m+2} = \sum_\ell a_\ell\,\Theta_{n+1,m+2} \quad (\ell \in \mathbb{Z} \bmod(m+3)\mathbb{Z}),$$

with

$$a_\ell = \sum_i q^{(m+2)(m+3)i^2},$$

where $i = j + [-(n+1) + 2\ell(m+2)]/[2(m+2)(m+3)]$ and $j \in \mathbb{Z}$. Note that

$$(m+2)(m+3)i^2 - \frac{(n+1)^2}{4(m+2)} = (m+2)(m+3)j^2$$

$$+ (-(n+1) + 2\ell(m+2))j + \ell^2 - \frac{(n+1+2\ell)^2}{4(m+3)}.$$

Hence

$$\Theta_{0,1}\Theta_{n+1,m+2}\, q^{-\frac{(n+1)^2}{4(m+2)}} = \sum_\ell b_\ell\,\Theta_{n+1+2\ell,m+3}\, q^{-\frac{(n+1+2\ell)^2}{4(m+3)}},$$

where

$$b_\ell = \prod_{j\in\mathbb{Z}} q^{(m+2)(m+3)j^2 + (-(n+1)+2\ell(m+2))j + \ell^2}.$$

Putting $\ell = -k$ and making the transformation $j \to -j$, we see that $b_{-k} = f_k^{(m,n)}$, so that

$$\Theta_{0,1}\Theta_{n+1,n+2}\, q^{-\frac{(n+1)^2}{4(m+2)}} = \sum_k f_k^{(m,n)}\Theta_{n+1-2k,m+3}\, q^{-\frac{(n+1-2k)^2}{4(m+3)}},$$

where $k \in \mathbb{Z} \bmod(m+3)\mathbb{Z}$. Similarly,

$$\Theta_{0,1}\Theta_{-n-1,m+2}\, q^{-\frac{(n+1)^2}{4(m+2)}} = \sum_k f_k^{(m,n)}\Theta_{-n-1+2k,m+3}\, q^{-\frac{(n+1-2k)^2}{4(m+3)}}$$

where $k \in \mathbb{Z} \bmod(m+3)\mathbb{Z}$.

Hence

$$\mathrm{ch}_d \, \mathrm{ch}_\lambda = \sum_k \frac{f_k^{(m,n)}}{\varphi(q)} \, \mathrm{ch}_{d+\lambda-k\alpha} \quad (k \in \mathbb{Z} \bmod (m+3)\mathbb{Z}). \qquad (11.23)$$

Now $d + \lambda - k\alpha = (m+1)d + \frac{1}{2}(n-2k)\alpha$; this corresponds to a non-negative integral linear combination of the fundamental weights ω_0 and ω_1 if and only if $k \in I$, where I was defined in (11.22b). We use the freedom in choosing the domain of definition of k in (11.23), i.e. of $\mathbb{Z} \bmod (m+3)\mathbb{Z}$, by taking k to be in the union of the sets I, J and K, where:

$$I = \left\{ k \in \mathbb{Z} \, \middle| \, -\frac{1}{2}(m+1-n) \leqslant k \leqslant \frac{1}{2}n \right\},$$

$$J = \left\{ k \in \mathbb{Z} \, \middle| \, \frac{1}{2}n + 1 \leqslant k \leqslant \frac{1}{2}(m+3+n) \right\},$$

$$K = \left\{ k \in \mathbb{Z} \, \middle| \, k = \frac{n+1}{2}, \quad k = \frac{m+n}{2} + 2 \right\}.$$

Note that K is nonempty only if n is odd, or if $m+n$ is even. The following symmetry property of the function $\mathrm{ch}_{d+\lambda-k\alpha}$ is easily checked:

$$\text{if } k \to n+1-k, \quad \text{then } \mathrm{ch}_{d+\lambda-k\alpha} \to -\mathrm{ch}_{d+\lambda-k\alpha}. \qquad (11.24)$$

An immediate corollary is that,

$$\mathrm{ch}_{d+\lambda-k\alpha} = 0 \quad \text{if } 2k = n+1 (\bmod m+3).$$

Hence, $\mathrm{ch}_{d+\lambda-k\alpha} = 0$ for $k \in K$. Then, since J maps onto I under the transformation $k \to n+1-k$, (11.23) becomes (11.22) on using (11.24). ∎

LECTURE 12

12.1. Preliminaries on $\hat{s\ell}_2$

In this lecture we shall use the Goddard-Kent-Olive (GKO) construction, described in Lecture 10, to construct unitary highest weight representations $V(c, h)$ of the Virasoro algebra for all pairs (c, h) given by the discrete series:

$$c = c_m \equiv 1 - \frac{6}{(m+2)(m+3)} \quad (m = 0, 1, 2, \ldots), \qquad (12.1a)$$

$$h = h_{r,s}^{(m)} \equiv \frac{[(m+3)r - (m+2)s]^2 - 1}{4(m+2)(m+3)} \quad (r, s \in \mathbb{N}, \ 1 \leqslant s \leqslant r \leqslant m+1). \qquad (12.1b)$$

In the course of the construction we shall also be able to prove Lemma 8.6 and thereby complete the proof of the Kac determinant formula. The unitarity of the discrete series was proved independently by Goddard-Kent-Olive [1986], Kac-Wakimoto [1986] and Tsuchiya-Kanie [1986]. This lecture follows closely the paper Kac-Wakimoto [1986].

We now summarize some properties of $s\ell_2$ and $\hat{s\ell}_2$ that we shall use. We choose the standard basis of $s\ell_2 : u_1 = e, u_2 = \alpha, u_3 = f$, where

$$e = \begin{bmatrix} 0 & 1 \\ 0 & 0 \end{bmatrix}, \quad \alpha = \begin{bmatrix} 1 & 0 \\ 0 & -1 \end{bmatrix}, \quad f = \begin{bmatrix} 0 & 0 \\ 1 & 0 \end{bmatrix}.$$

Then

$$\hat{n}_+ = \mathbb{C}e + \sum_{k>0} t^k s\ell_2 \,.$$

Defining the dual basis $\{u^i \mid i = 1, 2, 3\}$ by (10.2) with respect to the trace

123

form $(a \mid b) = \operatorname{tr} ab$, we have

$$u^1 = f, \quad u^2 = \alpha/2, \quad u^3 = e.$$

Then, by Lemma 10.1, the Casimir operator Ω_0 is given by

$$\Omega_0 = u_i u^i = ef + fe + \alpha^2/2 = \alpha^2/2 + \alpha + 2fe. \tag{12.2}$$

(Note that in (12.2) we have used the commutation relations to move the element e to the right.) We can compute the eigenvalue of the Casimir in its adjoint representation directly and we find that

$$g = 2, \tag{12.3}$$

which is in agreement with Table 10.1.

Let

$$P_+^0 = \left\{ md + \frac{1}{2}n\alpha \,\Big|\, m, n \in \mathbb{Z}_+, \ m \geqslant n \right\}; \quad P_+ = P_+^0 + \mathbb{R}M. \tag{12.4}$$

Recall from Lecture 9 that given $\lambda = md + 1/2n\alpha + rM \in P_+$ there exists a unique (up to equivalence) irreducible unitary highest weight representation of $\hat{s\ell}_2$ on a complex vector space $L(\lambda)$. This representation of $\hat{s\ell}_2$ remains irreducible and is independent of r when viewed as a representation of $\hat{s\ell}_2'$.

12.2.   A tensor product decomposition of some representations of $\hat{s\ell}_2$

We shall now consider the tensor product of two unitarizable highest weight representation of $\hat{s\ell}_2$, *viz.* $L(d)$ and $L(\lambda)$ where

$$\lambda = md + \frac{1}{2}n\alpha \in P_+^0. \tag{12.5}$$

The GKO construction gives us a unitary representation of Vir which commutes with $\hat{s\ell}_2'$. Hence the space $L(d) \otimes L(\lambda)$ can be reduced with respect to the direct sum of Vir and $\hat{s\ell}_2'$.

An essential tool in the following will be the character of the representations of $\hat{s\ell}_2$, which is computed from the Weyl-Kac character formula as we saw in Lecture 11. The usefulness of the character is due to its simple algebraic properties: the character of a tensor product of representations is the product of their characters while the character of a direct sum is the sum of their characters. Moreover, two representations are equivalent if and only if their characters are equal.

The characters of $L(d)$ and $L(\lambda)$ were denoted ch_d and ch_λ in Lecture 11. The character of $L(d) \otimes L(\lambda)$ is therefore $\mathrm{ch}_d\, \mathrm{ch}_\lambda$. We derived an important identity for this quantity in Proposition 11.4. We shall, however, have to do some more work to bring it to the form in which we need it:

Lemma 12.1. Let

$$
\left.
\begin{array}{lll}
r = n+1, & s = n+1-2k & \text{if } k \geqslant 0 \\
r = m-n+1, & s = m-n+2+2k & \text{if } k < 0
\end{array}
\right\}. \qquad (12.6)
$$

(Note that $1 \leqslant s \leqslant r \leqslant m+1$ since $m \geqslant n$ and $m, n \in \mathbb{Z}_+$.) Then

$$
\varphi(q)q^{-k^2}\psi_{m,n;k} \equiv q^{-k^2}(f_k^{(m,n)} - f_{n+1-k}^{(m,n)}) = A + B + C, \qquad (12.7a)
$$

where

$$
A = 1 - q^{rs} - q^{(m+2-r)(m-3-s)},
$$

$$
B = \sum_{j \in \mathbb{N}} q^{(m+2)(m+3)j^2 + ((m+3)r - (m+2)s)j}\left(1 - q^{2(m+2)sj + rs}\right), \qquad (12.7b)
$$

$$
C = \sum_{j \in \mathbb{N}} q^{(m+2)(m+3)j^2 - ((m+3)r - (m+2)s)j}\left(1 - q^{2(m+2)(m+3-s)j + (m+2-r)(m+3-s)}\right).
$$

Proof. A straightforward calculation using the definition (11.22c, d). ∎

Remark 12.1. Note from (12.7) that

$$
\psi_{m,n;k} = \frac{q^{k^2}}{\varphi(q)}\left(1 - q^{rs} - q^{(m+2-r)(m+3-s)} + \text{higher powers of } q\right). \qquad (12.8)
$$

∎

Proposition 12.1. Let $\lambda = md + \tfrac{1}{2}n\alpha$ $(m, n \in \mathbb{Z}_+;\ m \geqslant n)$. Then,

$$
\mathrm{ch}_d\, \mathrm{ch}_\lambda = \sum_{k \in I}\sum_{j \in \mathbb{Z}_+} \Delta_{m,n;k}^j\, \mathrm{ch}_{d+\lambda-k\alpha-jM} \qquad (12.9a)
$$

where

$$I = \left\{ k \in \mathbb{Z} \, \Big| \, -\frac{1}{2}(m+1-n) \leqslant k \leqslant \frac{1}{2}n \right\}. \qquad (12.9b)$$

The $\Delta^j_{m,n;k}$ are nonnegative integers defined by the expansion

$$\psi_{m,n;k} = \sum_{j \in \mathbb{Z}_+} \Delta^j_{m,n;k} \, q^j \,. \qquad (12.10)$$

The minimum value of j appearing with a nonzero coefficient in (12.10) is $j = k^2$, and $\Delta^{k^2}_{m,n;k} = 1$.

Proof. The fact that $\Delta^j_{m,n;k} \in \mathbb{Z}_+$ is clear from their representation-theoretical meaning explained below.

To prove the last statement of the proposition, if suffices to show that A, B and C in (12.7b) have each an expansion in powers of q with non-negative integer coefficients when multiplied by $\varphi(q)^{-1}$. Note that in B and C the term in brackets is always of the form $1 - q^i$ ($i > 0$). Hence on multiplying by $\varphi(q)^{-1} = \prod_{j=1}^{\infty}(1 - q^j)^{-1}$ one such factor cancels. The remainder will have an expansion of the required kind. It is straightforward to use the expansion (2.8a) to show by direct multiplication that $A/\varphi(q)$ has the required expansion.

Noting that the lowest power of q in (12.8) is q^{k^2}, this proves the last statement of the proposition. From (11.17) we see that

$$\mathrm{ch}_{\lambda - jM} = q^j \, \mathrm{ch}_\lambda \,.$$

Recalling Proposition 11.4, this completes the proof of Proposition 12.1. ∎

Proposition 12.1 has a very simple representation-theoretical meaning. Equation (12.9) expresses the decomposition of $L(d) \otimes L(\lambda)$ into a direct sum of unitary $\hat{s\ell}_2$ representations $L(d + \lambda - k\alpha - jM)$:

$$L(d) \otimes L(\lambda) = \sum_{k \in I} \sum_{j \in \mathbb{Z}_+} \Delta^j_{m,n;k} L(d + \lambda - k\alpha - jM)\,, \qquad (12.11)$$

and $\Delta^j_{m,n;k}$ is simply the multiplicity of the occurrence of $L(d+\lambda-k\alpha-jM)$ in this decomposition. ∎

12.3. Construction and unitarity of the discrete series representations of *Vir*

Let $U^{(j)}_{m,n;k}$ denote the space of highest weight vectors of $\hat{s\ell}_2$ in $L(d) \otimes L(\lambda)$ with highest weight $d + \lambda - k\alpha - jM \in P_+$. Then

$$\Delta^j_{m,n;k} = \dim U^{(j)}_{m,n;k} \,. \tag{12.12}$$

Clearly $U^{(j)}_{m,n;k}$ is an eigenspace of d with eigenvalue $(d \mid d + \lambda - k\alpha - jM) = -j$. Now define

$$U_{m,n;k} = \bigoplus_{j \in \mathbb{Z}_+} U^{(j)}_{m,n;k} \,. \tag{12.13}$$

It is clear that $U_{m,n;k}$ is the space of highest weight vectors of $\hat{s\ell}'_2$ in $L(d) \otimes L(\lambda)$ with highest weight $d + \lambda - k\alpha$. Now the GKO construction produces a representation of *Vir* which commutes with $\hat{s\ell}'_2$ and hence *Vir* maps $U_{m,n;k}$ into itself. The central charge of this representation of *Vir* in $U_{m,n;k}$ is given by (10.24) with $m' = 1$, $g = 2$ and $\dim \mathfrak{p} = 3$, which gives $c = c_m$ ($m \in \mathbb{Z}_+$), where c_m is given by (12.1a). The operator L_0 in this representation is given by (10.25) with

$$\Lambda = d, \quad \Lambda' = md + \frac{1}{2}n\alpha, \quad 2\rho = 4d + \alpha \,.$$

Thus

$$L_0 = \frac{n(n+2)}{4(m+2)} - \frac{\Omega}{2(m+3)} \,.$$

We deduce from (10.18) and (10.19):

$$\Omega = 2(M+2)d + \Omega_0 + 2 \sum_{j>0} u_i(-j)u^i(j) \,,$$

where $\Omega_0 = \alpha^2/2 + \alpha + 2fe$ from (12.2). Each $v \in U_{m,n;k}$ is a highest weight vector and hence is annihilated by the $u(j)$ with $j > 0$ and by e. Thus

$$\Omega(v) = (2d(M+2) + \alpha^2/2 + \alpha)\,(v) \quad \text{for } v \in U_{m,n;k} \,.$$

Since

$$d + \lambda - k\alpha = (m+1)d + \frac{n-2k}{2}\,\alpha = \tilde{\Lambda} \quad \text{(say)},$$

the level of $\tilde{\Lambda}$ is $m+1$, and $(\alpha \,|\, \tilde{\Lambda}) = n - 2k$, we obtain:

$$\Omega(v) = \left(2d(m+3) + \frac{(n-2k)^2}{2} + (n-2k)\right)v \quad \text{for } v \in U_{m,n;k}\,.$$

Thus in $U_{m,n;k}$ we have

$$L_0 = -d + \frac{n(n+2)}{4(m+2)} - \frac{(n-2k)(n-2k+2)}{4(m+3)}\,. \tag{12.14}$$

To determine the minimal eigenvalue of L_0 on $U_{m,n;k}$ we need to know the minimal eigenvalue of $(-d)$ on $U_{m,n;k}$. According to (12.13), $U_{m,n;k}$ is the direct sum of eigenspaces of $(-d)$: $U^{(j)}_{m,n;k}$ is an eigenspace of $(-d)$ with eigenvalue j. Hence we need to know the minimal value of j for which $\Delta^j_{m,n;k}$ is nonzero (see (12.12)). This was determined in Proposition 12.1 to be $j = k^2$. Thus the minimal eigenvalue of L_0 in the representation of Vir on $U_{m,n;k}$ is

$$h = k^2 + \frac{n(n+2)}{4(m+2)} - \frac{(n-2k)(n-2k+2)}{4(m+3)}\,. \tag{12.15}$$

Changing from the variables m, n, k to m, r, s by (12.6), we find that $h = h^{(m)}_{r,s}$ as given by (12.1b).

Denoting $U_{m,n;k}$ in the following by $U^{(m)}_{r,s}$, we have thus constructed a unitary representation of Vir on the space $U^{(m)}_{r,s}$ such that the central charge is c and the minimal eigenvalue of L_0 is h for every pair (c, h) in the discrete series (12.1a, b). All the eigenvalues of L_0 are from $h + \mathbb{Z}_+$. The highest component of the representation of Vir on $U^{(m)}_{r,s}$, i.e. the subrepresentation generated by the eigenvector of L_0 with the minimal eigenvalue $h = h^{(m)}_{r,s}$, is an irreducible unitary highest weight representation. This completes the proof that all representations $V(c, h)$ with (c, h) from the list (12.1a, b) are unitary.

From the above discussion we also obtain:

$$\operatorname{ch} V(c_m, h^{(m)}_{r,s}) \leqslant \operatorname{ch} U^{(m)}_{r,s}\,. \tag{12.16}$$

From (12.14) and (12.15) we see that $U^{(j)}_{m,n;k}$ is an eigenspace of L_0 with eigenvalue $(j - k^2) + h^{(m)}_{r,s}$. From (12.10) and (12.12) we have

$$\psi_{m,n;k} = \sum_{j \in \mathbb{Z}_+} \dim U^{(j)}_{m,n;k}\, q^j\,.$$

Thus $\psi_{m,n;k}$ is essentially the character of this representation of *Vir*. More precisely,

$$\operatorname{ch} U_{m,n;k} = q^{h_{r,s}^{(m)}} q^{-k^2} \psi_{m,n;k},$$

where $\psi_{m,n;k}$ is given by (12.7a, b). From (12.16) and Remark 12.1 we have:

$$\operatorname{ch} V(c_m, h_{r,s}^{(m)}) \leqslant \frac{q^{h_{r,s}^{(m)}}}{\varphi(q)} (1 - q^{rs} - q^{(m+2-r)(m+3-s)})$$

$$+ \text{ higher degree terms in } q. \qquad (12.17)$$

Remark 12.2. We can now interpret Proposition 11.4 as saying that for $\lambda = md + 1/2\, n\alpha \in P_+^0$ we have with respect to $\hat{s\ell}_2' \oplus Vir$:

$$L(d) \otimes L\left(md + \frac{1}{2} n\alpha\right) = \bigoplus_{\substack{0 \leqslant j \leqslant n \\ j \equiv n (\operatorname{mod} 2)}} L\left((m+1)d + \frac{1}{2} j\alpha\right) \otimes U_{n+1,j+1}^{(m)}$$

$$\bigoplus_{\substack{n+1 \leqslant j \leqslant m+1 \\ j \equiv n (\operatorname{mod} 2)}} L\left((m+1)d + \frac{1}{2} j\alpha\right) \otimes U_{m-n+1,m+2-j}^{(m)}. \qquad (12.18)$$

Keeping m fixed in (12.18) and varying n in the range $0 \leqslant n \leqslant m$, we see that on the right-hand side we get all $U_{r,s}^{(m)}$ satisfying $1 \leqslant s \leqslant r \leqslant m+1$, each occurring exactly once. It follows that with respect to *Vir*:

$$\left[L(d) \otimes \sum_{\substack{n \in \mathbb{Z}_+ \\ 0 \leqslant n \leqslant m}} L\left(md + \frac{1}{2} n\alpha\right) \right]^{\hat{n}_+} = \bigoplus_{\substack{r,s \in \mathbb{N} \\ 1 \leqslant s \leqslant r \leqslant m+1}} U_{r,s}^{(m)}, \qquad (12.19)$$

where the notation on the left-hand side means the subspace in the tensor product annihilated by $\hat{n}_+$. If we extend the sums in (12.19) over all possible values of $m \in \mathbb{Z}_+$, we see that every representation of the discrete series appears at least once in this space. ∎

We summarize below the results obtained:

Theorem 12.1. (a) The irreducible representation $V(c, h)$ of *Vir* is unitary for $c = c_m$, $h = h_{r,s}^{(m)}$ (given by (12.1a, b)), where $m, r, s \in \mathbb{Z}_+$, $1 \leqslant s \leqslant r \leqslant m+1$, and for $c \geqslant 1$, $h \geqslant 0$.

(b) With respect to Vir we have:

$$\left[L(d) \otimes \sum_{\substack{m,n\in\mathbb{Z}_+ \\ m\geqslant n}} L\left(md + \frac{1}{2}n\alpha\right)\right]^{\hat{n}_+} = \bigoplus_{\substack{m,r,s\in\mathbb{Z}_+ \\ 1\leqslant s\leqslant r\leqslant m+1}} U_{r,s}^{(m)},$$

where the highest component of $U_{r,s}^{(m)}$ is $V(c_m, h_{r,s}^{(m)})$.

(c) $\operatorname{ch} V(c_m, h_{r,s}^{(m)}) \leqslant \operatorname{ch} U_{r,s}^{(m)}$

$$= \frac{q^{h_{r,s}^{(m)}}}{\varphi(q)}(1 - q^{rs} - q^{(m+2-r)(m+3-s)} + B + C), \tag{12.20a}$$

where B and C are given by (12.7b).

Equivalently:

$$q^{-c_m/24}\operatorname{ch} V(c_m, h_{r,s}^{(m)}) \leqslant q^{-c_m/24}\operatorname{ch} U_{r,s}^{(m)}$$

$$= \frac{1}{\eta(\tau)}(\Theta_{r(m+3)-s(m+2),(m+2)(m+3)}(\tau,0,0)$$

$$- \Theta_{r(m+3)+s(m+2),(m+2)(m+3)}(\tau,0,0)), \tag{12.20b}$$

where

$$\eta(\tau) = q^{1/24}\varphi(q). \tag{12.20c}$$

(Here $1\leqslant s\leqslant r\leqslant m+1$, $m,r,s\in\mathbb{Z}_+$.) ∎

Remark 12.3. In fact, in (12.16), (12.17) and (12.20a, b) one has equality; in other words the representation of Vir on $U_{r,s}^{(m)}$ is irreducible and coincides with $V(c_m, h_{r,s}^{(m)})$. The most straightforward way of proving this is to calculate $\operatorname{ch} V(c_m, h_{r,s}^{(m)})$ and to notice that it coincides with the right-hand side of (12.20). The computation of $\operatorname{ch} V(c, h)$ consists of two steps, both of which are based on the Kac determinant formula. First, one finds all possible inclusions of Verma representations and the irreducible subfactors contained in $M(c, h)$; this was established by Kac [1978a] using the Jantzen filtration (see e.g. Kac-Kazhdan [1979]). Secondly, one shows that the irreducible subfactors of $M(c, h)$ occur with multiplicity one. This is the more difficult part, conjectured by Kac [1982] and proved by Feigin and Fuchs [1983a, b], [1984]. A simple proof of this fact has been given later by Astashkevich [1997]. The explicit formulae for all $\operatorname{ch} V(c, h)$ are easily derived from these facts: see Feigin-Fuchs [1983b], [1984]. Thus Theorem 12.1(b) provides a "model" for the discrete series of *Vir*, i.e. the space in which all the representations of the discrete series appear and exactly once. ∎

Remark 12.4. Formula (12.20b) shows that the functions $q^{-c_m/24}$ $\operatorname{ch} V(c_m, h_{r,s}^{(m)})$ are modular functions in τ (Im $\tau > 0$), where as before $q = \exp 2\pi i \tau$. Also, Remark 11.2 shows that the same is true for the functions $q^{-s_\lambda} \operatorname{tr}_{L(\lambda)} q^{-d}$. This observation has played an important role in subsequent developments in representation theory, as well as in the theory of 2-dimensional statistical models and string theory (see Kac-Peterson [1984a], Kac-Wakimoto [1987], [1988], Capelli-Itzykson-Zuber [1987], Gepner-Witten [1986], Zhu [1996], and references there). ∎

12.4. Completion of the proof of the Kac determinant formula

We shall now show how the inequality for $\operatorname{ch} V(c_m, h_{r,s}^{(m)})$ given by (12.17) can be used to prove Lemma 8.6, thereby completing the proof of the Kac determinant formula discussed in Lecture 8. We recall from (3.15) that the character of the Verma representation $M(c_m, h_{r,s}^{(m)})$ is $q^{h_{r,s}^{(m)}}/\varphi(q)$. Let $r' = m + 2 - r$, $s' = m + 3 - s$ so that $s' > r'$. Note from (12.1b) that $h_{r,s}^{(m)} = h_{r',s'}^{(m)}$. From the occurrence in (12.17) of the lowest powers of q, viz. q^{rs} and $q^{r's'}$, with negative coefficients, we can deduce that $J(c_m, h_{r,s}^{(m)})$ (the maximal proper subrepresentation of $M(c_m, h_{r,s}^{(m)})$) has a nonzero component at each level $n \geqslant \min(rs, r's')$. Thus $\det_n(c, h)$ has a zero at $h = h_{r,s}^{(m)}$ ($1 \leqslant s \leqslant r \leqslant m + 1$) for $n \geqslant \min(rs, r's')$. Thus $\det_n(c_m, h)$ has a zero at $h = h_{r,s}^{(m)}$ for all pairs (r, s) satisfying $1 \leqslant r, s \leqslant m+1$ and $n \geqslant rs$. In Theorem 8.1 we defined $\varphi_{r,s}(c, h) = (h - h_{r,s})(h - h_{s,r})$ for $r \neq s$ and $\varphi_{r,r}(c, h) = h - h_{r,r}$. Viewed as a polynomial in the two variables c, h, it is clear that $\varphi_{r,s}(c, h)$ is irreducible (over the complex field), i.e. it cannot be written as the product of linear factors (for $r \neq s$). Now, $\det_n(c, h)$ vanishes at an infinite number of points $(c_m, h_{r,s}^{(m)})$ of the irreducible curve $\varphi_{r,s}(c, h) = 0$ for $n \geqslant rs$. Hence $\det_n(c, h)$ vanishes at all points of $\varphi_{r,s}(c, h) = 0$ for $n \geqslant rs$ and therefore is divisible by $\varphi_{r,s}(c, h)$ when $rs \leqslant n$. The proof of Lemma 8.6, and hence of the Kac determinant formula, is complete.

Remark 12.5. There have been several published proofs of the Kac determinant formula: see Feigin-Fuchs [1982], Thorn [1984] etc. The proof given here is simpler, more elegant and works in the super case as well (see Kac-Wakimoto [1986]).   ∎

Remark 12.6. Let $h = h_{r,s}^{(m)}$ be one of the members of the list (12.1a, b) and suppose that there is no h' from this list (for the same m) with $h' - h$ a positive integer. Then in (12.16), (12.17) and (12.20) we have equality. This follows from the fact that the lowest eigenspace of d appearing in (12.13) has dimension 1 ($\Delta_{m,n;k}^j = 1$ for $j = k^2$) so that if $V(c, h')$ were contained in $U_{r,s}^{(m)}$, $h' - h$ must be a positive integer (see the remarks after (12.15)). In that case, by our hypothesis, (c_m, h') does not belong to the discrete series (12.1a, b) and hence cannot be unitary by the theorem of Friedan-Qiu-Shenker (FQS) [1984; 1986]; however, by the argument of Proposition 3.1, $U_{r,s}^{(m)}$ is a direct sum of unitary representations of *Vir*. We conclude that equality holds in (12.16), (12.17) and (12.20). The above condition holds in most, but not all, cases. In particular, it holds for $m = 1, 2$.   ∎

12.5.   On non-unitarity in the region $0 \leqslant c < 1, h \geqslant 0$

Recall that, according to the FQS theorem, all points in the *critical region* $0 \leqslant c < 1, h \geqslant 0$ not belonging to the discrete series (12.1) correspond to non-unitary representations of *Vir*. This together with Propositions 3.5 and 8.2(a) and Theorem 12.1(a), gives a complete classification of unitary highest weight representations of *Vir*: either $c \geqslant 1$ and $h \geqslant 0$ or $(c, h) \in$ discrete series (12.1).

In the remainder of this last lecture we shall discuss non-unitarity in the critical region. We shall obtain here only some partial results, which are, however, most important for applications (as in Proposition 3.8 or Remark 12.6, for example). The proof of the general result is more involved. See Friedan-Qiu-Shenker [1986] or Langlands [1986].

We call the n-th *ghost number*, and denote it by $g_n(c, h)$, the number of negative eigenvalues of the matrix of the Hermitian contravariant form $\langle \cdot \, | \, \cdot \rangle$ restricted to the n-th level of a highest weight representation with highest weight (c, h). Note that $g_n(c, h)$ depends only on n, c and h (in particular, it is the same for $V(c, h)$ and $M(c, h)$) and that $V(c, h)$ is unitary if $g_n(c, h) = 0$ for all $n \in \mathbb{Z}_+$.

Lemma 12.2. If $h \geqslant 0$ and $n \in \mathbb{Z}_+$, then

$$g_n(c, h) \leqslant g_{n+1}(c, h).$$

Proof. Let $e = d_1$, $\alpha = -2d_0$, $f = -d_{-1}$. These elements form a standard basis of an $s\ell_2$ subalgebra of Vir, which we denote by $\mathfrak{a}$. Consider $V(c,h)$ as a representation of $\mathfrak{a}$. Given $v \in V(c,h)_{h+n}^e$ (i.e. $\alpha(v) = -2(h+n)v$, $e(v) = 0$), we have for $\lambda = -2(h+n)$ (see e.g. the proof of Lemma 8.1):

$$\langle f^k(v) \mid f^k(v) \rangle = k!(-\lambda)(-\lambda+1)\ldots(-\lambda+k-1)\langle v \mid v \rangle, \quad (12.21)$$

$$ef^k(v) = -k(k-1+2(h+n))\, f^{k-1}(v). \quad (12.22)$$

It follows that unless $v \in V(c,h)_h$ and $h = 0$ (then $\mathbb{C}v$ is $\mathfrak{a}$-invariant), the subspace $T_v = \sum_{k \geq 0} \mathbb{C}f^k(v)$ is a space of an irreducible representation of $\mathfrak{a}$ whose intersection with each $V(c,h)_{h+N}$ is 1-dimensional for $N \geq n$ (by (12.22)). Also, by (12.21), the Hermitian form $\langle \cdot \mid \cdot \rangle$ restricted to T_v is positive definite, negative definite or zero according as $\langle v \mid v \rangle > 0$, < 0 or $= 0$ respectively.

Using the Casimir operator of $\mathfrak{a}$, it is easy to see that $V(c,h)$ decomposes in a direct sum of subspaces T_v, where the vectors v are some eigenvectors of L_0 with nonzero $\langle v \mid v \rangle$. This completes the proof of the lemma. ∎

Proposition 12.2. Let

$$\mathscr{D}_j = \{(c,h) \mid 0 \leq c < 1,\ h \geq 0 \ \text{ and }\ \varphi_{j,1}(c,h) < 0\},$$

where $\varphi_{r,s}$ are defined in (8.13). Then for every $(c,h) \in \cup_{j \geq 2} \mathscr{D}_j$, the representation $V(c,h)$ is not unitary.

Proof. We prove by induction on $n \geq 2$ that for (c,h) from the region $\mathscr{D}^{(n)} = \cup_{j=2}^n \mathscr{D}_j$, one has: $g_n(c,h) \geq 1$. The case $n = 2$ has been discussed in Lecture 8. Suppose that the statement is true for $n-1$. Then $g_n(c,h) \geq 1$ in $\mathscr{D}^{(n-1)}$ by Lemma 12.2. Let $(c,h) \in \mathscr{D}_n \backslash \mathscr{D}^{(n-1)}$; it is easy to see (using formulas (8.13) or (8.15)) that then $\varphi_{r,s}(c,h) > 0$ for all r,s such that $rs \leq n$, $(r,s) \neq (n,1)$. Hence, by the Kac determinant formula, $\det_n(c,h) < 0$, proving that $g_n(c,h) \geq 1$ in $\mathscr{D}^{(n)}$ except for the curve $\gamma = \mathscr{D}_n \cap \{(c,h) \mid \varphi_{n-1,1}(c,h) = 0\}$ (see Figure 12.1). The curve γ is tackled as follows. Recall that in the region $\mathscr{D}_n \cap \mathscr{D}_{n-1}$, $g_n(c,h) \geq 1$. But $\det_n(c,h) = \varphi_n(c,h)\varphi_{n-1}(c,h) \times$ (the rest) > 0 in this region since the first two factors are negative and the rest are easily seen to be positive. Hence we conclude that

$$g_n(c,h) \geq 2 \quad \text{for}\ (c,h) \in \mathscr{D}_n \cap \mathscr{D}_{n-1}. \quad (12.23)$$

But the multiplicity of a zero of $\det_n(c, h)$ (as a polynomial in h with fixed c) is 1 for $(c, h) \in \gamma$. This shows that $g_n(c, h) \geqslant 1$ on γ. ∎

Corollary 12.1. If $V(1/2, h)$ is unitary, then $h = 0$, $1/2$, or $1/16$.

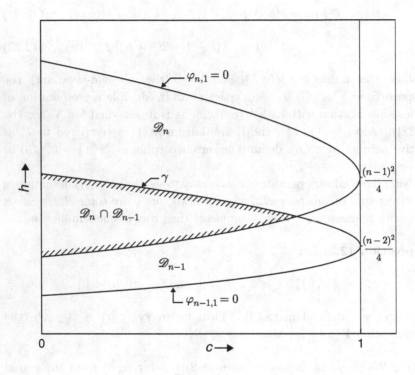

Figure 12.1

Proof. For $c = 1/2$, Proposition 12.2 eliminates all points except for $h = 1/2$ and $0 \leqslant h \leqslant 1/16$. But $\varphi_{2,2}(1/2, h) < 0$, and hence $\det_4(1/2, h) < 0$, for $0 < h < 1/16$. ∎

Remark 12.7. A similar argument shows that the unitarity of $V(c_m, h)$, where c_m is given by (12.1a) implies that $h = h_{r,s}^{(m)}$ given by (12.1b). ∎

LECTURE 13

13.1. Formal distributions

A quantum field is an operator-valued distribution on space-time. This is a key idea of modern quantum field theory, introduced by Wightman in order to incorporate mathematically the deep analysis of the notion of a quantum field by Bohr, Rosenfeld, Landau, and Peierls. In conformal quantum field theory one can often express a quantum field in terms of "chiral" quantum fields, which are a special kind of vector-valued formal distributions.

Definition 13.1. Let U be a vector space over $\mathbb{C}$. A U-valued formal distribution is an expression of the form

$$a(z) = \sum_{n \in \mathbb{Z}} a_n z^n, \quad a_n \in U,$$

where z is an indeterminate.

Formal distributions form a complex vector space denoted by $U[[z, z^{-1}]]$ or $U[[z^{\pm 1}]]$. Moreover, $U[[z^{\pm 1}]]$ is a module over the algebra of Laurent polynomials in z since any formal distribution $a(z)$ can be multiplied by a power of z. However, in general $a(z)$ cannot be multiplied by an element of $\mathbb{C}[[z^{\pm 1}]]$, or even by a formal power series in z. (Example: $\sum_{n \in \mathbb{Z}} z^n$ cannot be multiplied by $\sum_{n \geq 0} z^n$.)

Define the *residue* of $a(z)$ by

$$\operatorname{Res}_z a(z) = a_{-1}.$$

This linear map induces a non-degenerate pairing $U[[z, z^{-1}]] \times \mathbb{C}[z, z^{-1}] \longrightarrow U$, given by

$$\langle a, f \rangle = \text{Res}_z \, f(z) \, a(z),$$

for $f(z) \in \mathbb{C}[z, z^{-1}]$, $a(z) \in U[[z, z^{-1}]]$. Each U-valued formal distribution $a(z)$ defines a U-valued $\mathbb{C}$-linear function on the space $\mathbb{C}[z, z^{-1}]$ of "test functions":

$$a(z)(f) = \langle a, f \rangle. \tag{13.1}$$

It is easy to see that by (13.1), the space of U-valued formal distributions is identified with the space of U-valued linear functions on $\mathbb{C}[z, z^{-1}]$, which explains the term "formal distribution".

The derivation ∂_z is an endomorphism of $U[[z^{\pm 1}]]$ that is defined in the obvious way. It is clear that for any U-valued formal distribution $a(z)$

$$\text{Res}_z \, (\partial_z \, a(z)) = 0. \tag{13.2}$$

One often writes a formal distribution $a(z)$ in the form

$$a(z) = \sum_{n \in \mathbb{Z}} a_{(n)} z^{-n-1}, \tag{13.3}$$

since for this indexing of coefficients one has: $a_{(n)} = \text{Res}_z \, z^n a(z)$. Then we have

$$\partial_z \, a(z) = -\sum_{n \in \mathbb{Z}} n a_{(n-1)} z^{-n-1}. \tag{13.4}$$

A U-valued formal distribution in two indeterminates z and w is an expression of the form

$$a(z, w) = \sum_{m,n \in \mathbb{Z}} a_{m,n} z^m w^n, \quad a_{m,n} \in U.$$

These formal distributions form a complex vector space, denoted by $U[[z^{\pm 1}, w^{\pm 1}]]$, which, similarly to (13.1), can be identified with the space of linear functions on the space of "test functions" $\mathbb{C}[z^{\pm 1}, w^{\pm 1}]$. The extension to more variables is evident.

Definition 13.2. A U-valued formal distribution $a(z, w)$ is said to be *local* if

$$(z - w)^N a(z, w) = 0 \text{ for some } N \in \mathbb{Z}_+.$$

The *formal delta function* is an important example of a formal distribution of two variables. It is defined by

$$\delta(z, w) = z^{-1} \sum_{n \in \mathbb{Z}} \left(\frac{w}{z} \right)^n. \tag{13.5}$$

By abuse of notation, this $\mathbb{C}$-valued formal distribution is traditionally denoted by $\delta(z - w)$, but we shall denote it instead as $\delta(z, w)$. Note that

$$\text{Res}_z \ \delta(z, w) = \text{Res}_w \ \delta(z, w) = 1. \tag{13.6}$$

The formal delta function is a local formal distribution, since, as will be shown in Proposition 13.1, $(z - w)\delta(z, w) = 0$. But first we clarify the relations between formal distributions and formal Laurent series expansions of rational functions.

Let $\mathcal{R}$ be the algebra of rational functions of variables z and w with poles only at $z = 0$, $w = 0$, or $|z| = |w|$. Any element $R(z, w)$ of the algebra $\mathcal{R}$ has an expansion in powers of w/z in the region $|z| > |w|$, and in powers of z/w in the region $|w| > |z|$. Denote these expansions as $\iota_{z,w} \ R(z, w)$ and $\iota_{w,z} \ R(z, w)$ respectively. The operators $\iota_{z,w}$ and $\iota_{w,z}$ are homomorphisms of algebras

$$\iota_{z,w} : \mathcal{R} \longrightarrow \mathbb{C}[z^{\pm 1}, w^{\pm 1}] \left[\left[\frac{w}{z} \right] \right], \quad \iota_{w,z} : \mathcal{R} \longrightarrow \mathbb{C}[z^{\pm 1}, w^{\pm 1}] \left[\left[\frac{z}{w} \right] \right].$$

In particular,

$$\iota_{z,w} \frac{1}{z - w} = \frac{1}{z} \sum_{n \geq 0} \left(\frac{w}{z} \right)^n, \tag{13.7}$$

$$\iota_{w,z} \frac{1}{z - w} = -\frac{1}{w} \sum_{n \geq 0} \left(\frac{z}{w} \right)^n = -\frac{1}{z} \sum_{n < 0} \left(\frac{w}{z} \right)^n. \tag{13.8}$$

The algebras $\mathbb{C}[z^{\pm 1}, w^{\pm 1}] \left[\left[\frac{w}{z} \right] \right]$ and $\mathbb{C}[z^{\pm 1}, w^{\pm 1}] \left[\left[\frac{z}{w} \right] \right]$ are subspaces of the vector space $\mathbb{C}[[z^{\pm 1}, w^{\pm 1}]]$, so, by the definition of the formal delta function, we can write in $\mathbb{C}[[z^{\pm 1}, w^{\pm 1}]]$:

$$\delta(z, w) = \iota_{z,w} \frac{1}{z - w} - \iota_{w,z} \frac{1}{z - w}. \tag{13.9}$$

Remark 13.1. The homomorphisms $\iota_{z,w}$ and $\iota_{w,z}$ obviously commute with multiplication by elements of $\mathbb{C}[z^{\pm 1}, w^{\pm 1}]$. They also commute with the derivations ∂_z and ∂_w. Indeed, for a rational function $R(z, w) = (z - w)^k$ we have

$$\partial_z \iota_{z,w} (z - w)^k = \partial_z \sum_{n \geq 0} \binom{k}{n} z^{k-n}(-w)^n = \sum_{n \geq 0} \binom{k}{n}(k - n)z^{k-n-1}(-w)^n$$

$$= k \sum_{n \geq 0} \binom{k - 1}{n} z^{k-n-1}(-w)^n$$

$$= \iota_{z,w} \, k(z - w)^{k-1} = \iota_{z,w} \, \partial_z(z - w)^k.$$

The other cases are proved similarly.   ∎

Differentiating both sides of (13.9) and (13.5) j times, we deduce the following formula:

$$\partial_w^j \, \delta(z, w)/j! = \iota_{z,w} \frac{1}{(z - w)^{j+1}} - \iota_{w,z} \frac{1}{(z - w)^{j+1}}$$

$$= \sum_{m \in \mathbb{Z}} \binom{m}{j} z^{-m-1} w^{m-j}. \tag{13.10}$$

It is important to observe that we can multiply a $\mathbb{C}$-valued formal distribution $b(z, w)$ by an arbitrary U-valued formal distribution $c(z)$, provided that the coefficients of w^n, $n \in \mathbb{Z}$, in $b(z, w)$ are Laurent polynomials in z, i.e. $b(z, w) \in \mathbb{C}[z^{\pm 1}][[w^{\pm 1}]]$. Note that the formal delta function and its derivatives do satisfy this property, hence their product with an arbitrary U-valued formal distribution in z or in w is well defined.

In the following proposition we collect the properties of the formal delta function that will be used in the sequel.

Proposition 13.1.

(a) $(z - w)^m \, \partial_w^n \, \delta(z, w) \;\; = \;\; 0 \quad$ if $m > n$.

(b) $(z - w) \dfrac{\partial_w^n}{n!} \, \delta(z, w) \;\; = \;\; \dfrac{\partial_w^{n-1}}{(n - 1)!} \, \delta(z, w)$ if $n \geq 1$.

(c) $\qquad\qquad\qquad \delta(z,w) \;=\; \delta(w,z).$

(d) $\qquad\qquad \partial_z\,\delta(z,w) \;=\; -\,\partial_w\,\delta(z,w).$

(e) $\qquad\quad a(z)\,\delta(z,w) \;=\; a(w)\,\delta(z,w) \quad \text{for } a(z) \in U[[z^{\pm1}]].$

(f) $\qquad \mathrm{Res}_z\, a(z)\,\delta(z,w) \;=\; a(w).$

(g) $\quad e^{\lambda(z-w)}\,\partial_w^n\,\delta(z,w) \;=\; (\lambda+\partial_w)^n\,\delta(z,w) \text{ if } n \geq 0.$

Proof. Properties (a)-(d) follow immediately from Remark 13.1 and formula (13.10).

Part (a) implies that $(z-w)\,\delta(z,w) = 0$, so $z^n\,\delta(z,w) = w^n\,\delta(z,w)$. Hence (e) follows. The property (f) is an immediate consequence of property (e) and formula (13.6).

In order to prove (g), notice that $e^{\lambda(z-w)}\partial_w e^{-\lambda(z-w)} = \partial_w + \lambda$, which is easily checked by applying both sides to $f(z,w) \in \mathbb{C}[[z^{\pm}, w^{\pm}]][[\lambda]]$. Raising both sides to n-th power, we obtain: $e^{\lambda(z-w)}\partial_w^n e^{-\lambda(z-w)} = (\partial_w + \lambda)^n$. Now we apply both sides of this equality to $\delta(z,w)$ and use that $e^{\lambda(z-w)}\delta(z,w) = \delta(z,w)$ (since $(z-w)\delta(z,w) = 0$), to get (g). ∎

Remark 13.2. From part (a) of Proposition 13.1 it follows immediately that any U-valued formal distribution of the form

$$a(z,w) \;=\; \sum_{j=0}^{N} a_j(w)\,\partial_w^j \delta(z,w), \quad \text{where} \quad a_j(w) \in U[[w^{\pm1}]] \quad \text{and } N \in \mathbb{Z}_+,$$

(13.11)

is local. ∎

In the proof of the important Decomposition Theorem below and further on we shall need the following simple lemma:

Lemma 13.1. Let $b(z,w)$ be a U-valued formal distribution in z and w of the form $b(z,w) = \sum_{k\geq 0} b_k(w)z^k$, where $b_k(w) \in U[[w^{\pm1}]]$. Suppose that $(z-w)^N b(z,w) = 0$ for some $N \in \mathbb{Z}_+$. Then $b(z,w) = 0$.

Proof. Suppose that not all of the $b_k(w)$ vanish, and take the least nonnegative integer k for which $b_k(w) \neq 0$. Then in the power series expansion in z of $(z-w)^N b(z,w)$ the coefficient of z^k is $(-w)^N b_k(w) \neq 0$, and we get a contradiction. ∎

Theorem 13.1. (Decomposition Theorem) Let $a(z, w)$ be a U-valued local formal distribution. Then it can be decomposed uniquely as a finite linear combination of derivatives of $\delta(z, w)$ with coefficients in $U[[w^{\pm 1}]]$:

$$a(z, w) = \sum_{j=0}^{N} c^j(w) \, \partial_w^j \, \delta(z, w)/j! \quad (N \in \mathbb{Z}_+). \qquad (13.12)$$

The coefficients $c^j(w)$ are given by

$$c^j(w) = \mathrm{Res}_z \, (z - w)^j \, a(z, w). \qquad (13.13)$$

Proof. If the decomposition

$$a(z, w) = \sum_{j=0}^{N} c^j(w) \, \partial_w^j \, \delta(z, w)/j!$$

exists, the coefficients $c^j(w)$ of that decomposition are uniquely determined and are given by the formula (13.13). Indeed, by parts (a), (b) and (f) of Proposition 13.1,

$$\mathrm{Res}_z \, (z - w)^n \, a(z, w) = \mathrm{Res}_z \, (z - w)^n \sum_{j=0}^{N} c^j(w) \, \partial_w^j \, \delta(z, w)/j! = c^n(w).$$

Let N be such that $(z - w)^N \, a(z, w) = 0$. Define a formal distribution $a_1(z, w)$ by the formula

$$a_1(z, w) = \sum_{j=0}^{N} c^j(w) \, \partial_w^j \, \delta(z, w)/j!, \qquad (13.14)$$

where $c^j(w)$ is given by (13.13).

The formal distribution $a_1(z, w)$ is local by Remark 13.2, so $a(z, w) - a_1(z, w)$ is also local. For any $k \geq 0$ the residue by z of the formal distribution $(z - w)^k (a(z, w) - a_1(z, w))$ is zero. Indeed, by Proposition 13.1(b), we have

$$\mathrm{Res}_z \, (z - w)^k \, (a(z, w) - a_1(z, w))$$

$$= \mathrm{Res}_z \left((z - w)^k a(z, w) - \sum_{j \geq k} c^j(w) \frac{\partial_w^{j-k}}{(j - k)!} \delta(z, w) \right).$$

From (13.13) and (13.6), the last expression is

$$c^k(w) - \sum_{j \geq k} c^j(w) \, \frac{\partial_w^{j-k}}{(j-k)!} \mathrm{Res}_z \, \delta(z,w) = c^k(w) - c^k(w) = 0.$$

Thus,

$$\mathrm{Res}_z \left((z-w)^k \left(a(z,w) - a_1(z,w) \right) \right) = 0 \quad \text{for any } \; k \geq 0,$$

and in the expansion of $a(z,w) - a_1(z,w) = \sum_{k \in \mathbb{Z}} b_k(w) z^k$ the coefficients $b_k(w)$ are 0 for $k < 0$.

Since $b(z,w) := a(z,w) - a_1(z,w)$ is a local formal distribution, there exists $M \in \mathbb{Z}_+$, such that $(z-w)^M b(z,w) = 0$. Applying Lemma 13.1, we obtain $a(z,w) = a_1(z,w)$. ∎

13.2. Local pairs of formal distributions

Consider formal distributions $a(z) = \sum_{n \in \mathbb{Z}} a_n z^n$ and $b(z) = \sum_{n \in \mathbb{Z}} b_n z^n$ with values in a Lie algebra $\mathfrak{g}$. Their commutator is a formal $\mathfrak{g}$-valued distribution in two variables, defined by

$$[a(z), b(w)] = \sum_{m,n \in \mathbb{Z}} [a_m, b_n] \, z^m w^n.$$

Definition 13.3. For any $n \in \mathbb{Z}_+$, the *n-th product* of $\mathfrak{g}$-valued formal distributions $a(w)$ and $b(w)$ is a formal $\mathfrak{g}$-valued distribution in w defined by

$$a(w)_{(n)} b(w) = \mathrm{Res}_z \, (z - w)^n [a(z), b(w)]. \tag{13.15}$$

We will often abbreviate a for $a(w)$, b for $b(w)$, ∂a for $\partial_w \, a(w)$, and $a_{(n)} b$ for the n-th product of $a(w)$ and $b(w)$.

Proposition 13.2. The operator ∂_w is a derivation of the n-th product for any $n \in \mathbb{Z}_+$.

Proof. We have

$$\partial_w \left(a(w)_{(n)} b(w) \right) = \mathrm{Res}_z \left((\partial_w \, (z-w)^n) [a(z), b(w)] \right) + a(w)_{(n)} \partial_w \, b(w).$$

Hence we need to check that $\text{Res}_z \left((\partial_w \ (z \ - \ w)^n)[a(z), b(w)] \right) = (\partial_w \ a(w))_{(n)} b(w)$, or that

$$\text{Res}_z \ (-(\partial_w \ (z - w)^n)[a(z), b(w)] + (z - w)^n[\partial_z \ a(z), b(w)]) =$$
$$\text{Res}_z \ ((\partial_z \ (z - w)^n)[a(z), b(w)] + (z - w)^n[\partial_z \ a(z), b(w)]) = 0.$$

But the left-hand side of the last line is $\text{Res}_z \ \partial_z \left((z - w)^n[a(z), b(w)] \right)$, hence zero by (13.2).   ∎

Definition 13.4. A pair (a, b) of $\mathfrak{g}$-valued formal distributions in one variable is said to be *local* if the formal distribution $[a(z), b(w)]$ in two variables is local. A collection of $\mathfrak{g}$-valued formal distributions, for which all pairs are local, is called a *local family*.

The Decomposition Theorem 13.1 immediately implies the following proposition.

Proposition 13.3. (Operator Product Expansion) Let (a, b) be a local pair of $\mathfrak{g}$-valued formal distributions. Then

$$[a(z) \, , \, b(w)] = \sum_{j=0}^{N} (a(w)_{(j)} b(w)) \ \partial_w^j \ \delta(z, w)/j! \quad \text{for some } N \in \mathbb{Z}_+.$$

$$(13.16)$$

   ∎

Formula (13.16) is called *(the singular part of) the Operator Product Expansion (OPE)*.

Write the $\mathfrak{g}$-valued formal distributions $a(z)$, $b(z)$ in the form (13.3). Using (13.10), substitute $\partial_w^j \ \delta(z, w)/j!$ in (13.16) by $\sum_m \binom{m}{j} z^{-m-1} w^{m-j}$ and compare the coefficients of $z^{-m-1} w^{-n-1}$ on both sides. Then we see that OPE (13.16) for the formal distributions a and b is equivalent to the following commutation relations of their coefficients:

$$[a_{(m)} \, , \, b_{(n)}] = \sum_{j \geq 0} \binom{m}{j} \left(a_{(j)} b \right)_{(m+n-j)}. \qquad (13.17)$$

By (13.4), we also have

$$(\partial a)_{(n)} = -n a_{(n-1)}. \qquad (13.18)$$

The rules (13.17) and (13.18) allow one to convert commutator formulas into OPE and vice versa, as will be demonstrated further on.

13.3. Formal Fourier transform

Definition 13.5. The *formal Fourier transform* $\mathcal{F}^\lambda_{z,w}$ of a U-valued formal distribution $a(z,w)$ is defined by

$$\mathcal{F}^\lambda_{z,w}\, a(z,w) = \operatorname{Res}_z e^{\lambda(z-w)}\, a(z,w). \tag{13.19}$$

Note that the product $e^{\lambda(z-w)} a(z,w)$ makes sense, since the coefficient of λ^n is a product of a polynomial in z and w and of $a(z,w)$. By definition, $\mathcal{F}^\lambda_{z,w}\, a(z,w)$ is the following formal power series in λ with coefficients in $U[[w^{\pm 1}]]$:

$$\mathcal{F}^\lambda_{z,w}\, a(z,w) = \sum_{j \geq 0} \frac{\lambda^j}{j!}\, c^j(w), \tag{13.20}$$

where the coefficients $c^j(w)$ are given by (13.13).

Proposition 13.4. The following are the main properties of the formal Fourier transform:

(a) $\mathcal{F}^\lambda_{z,w}\, \partial_z\, a(z,w) = -\lambda\, \mathcal{F}^\lambda_{z,w}\, a(z,w) = [\partial_w,\, \mathcal{F}^\lambda_{z,w}]\, a(z,w).$

(b) $\mathcal{F}^\lambda_{z,w}\, a(w,z) = \mathcal{F}^{-\lambda-\partial_w}_{z,w}\, a(z,w)$ if $a(z,w)$ is local, where the right-hand side is interpreted as the right-hand side of (13.20) with λ replaced by $-\lambda - \partial_w$.

(c) $\mathcal{F}^\lambda_{z,w}\, \mathcal{F}^\mu_{x,w}\, a(z,w,x) = \mathcal{F}^{\lambda+\mu}_{x,w}\, \mathcal{F}^\lambda_{z,x}\, a(z,w,x)$, where $a(z,w,x)$ is a formal U-valued distribution in three indeterminates.

Proof. (a) By (13.2) we have:

$$\mathcal{F}^\lambda_{z,w}\, (\partial_z\, a(z,w)) = \operatorname{Res}_z e^{\lambda(z-w)} \partial_z\, a(z,w) = -\operatorname{Res}_z \left(\partial_z\, e^{\lambda(z-w)} \right) a(z,w)$$

$$= -\lambda \operatorname{Res}_z e^{\lambda(z-w)} a(z,w) = -\lambda\, \mathcal{F}^\lambda_{z,w}\, a(z,w).$$

The second equality of (a) is a straightforward calculation.

(b) Since $a(z, w)$ is local, by the Decomposition Theorem 13.1, it suffices to consider $a(z, w) = c(w) \, \partial_w^j \, \delta(z, w)$. Then $a(w, z) = c(z) \partial_z^j \delta(z, w)$ and by Proposition 13.1,

$$
\begin{aligned}
\mathfrak{F}_{z,w}^\lambda \, a(w, z) &= \mathrm{Res}_z \, e^{\lambda(z-w)} c(z) \partial_z^j \, \delta(z, w) \\
&= \mathrm{Res}_z \, c(z) e^{\lambda(z-w)} (-\partial_w)^j \, \delta(z, w) \\
&= \mathrm{Res}_z \, c(z) (-\lambda - \partial_w)^j \, \delta(z, w) \\
&= \mathrm{Res}_z \, (-\lambda - \partial_w)^j \, c(w) \delta(z, w) \\
&= (-\lambda - \partial_w)^j c(w) = \mathfrak{F}_{z,w}^{-\lambda - \partial_w} \, a(z, w).
\end{aligned}
$$

(c) The composition of the formal Fourier transforms on the left-hand side of part (c) is a $\mathbb{C}$-linear map from $U[[z^{\pm 1}, w^{\pm 1}, x^{\pm 1}]]$ to $U[[w^{\pm 1}]] \, [[\lambda, \mu]]$. The property (c) immediately follows from the equality $e^{\lambda(z-w)+\mu(x-w)} = e^{(\lambda+\mu)(x-w)} e^{\lambda(z-x)}$. ∎

13.4. Lambda-bracket of local formal distributions

Definition 13.6. Let (a, b) be a local pair of $\mathfrak{g}$-valued formal distributions. The λ-*bracket* of a and b is defined by

$$
[a_\lambda b] = \sum_{n \in \mathbb{Z}_+} \frac{\lambda^n}{n!} \, (a_{\,(n)} \, b). \tag{13.21}
$$

It is polynomial in λ due to locality. By the OPE (13.16) and by (13.20), one can see that

$$
[a_\lambda b] = \mathfrak{F}_{z,w}^\lambda \, [a(z), \, b(w)]. \tag{13.22}
$$

Proposition 13.5. The λ-bracket satisfies the following properties:

(Sesquilinearity) $[\partial a_\lambda b] = -\lambda[a_\lambda b], \quad [a_\lambda \partial b] = (\lambda + \partial)[a_\lambda b];$ (13.23)

(Skewcommutativity) $[b_\lambda a] = -[a_{-\lambda-\partial} \, b]$ (if a, b is a local pair);   (13.24)

(Jacobi identity) $[a_\lambda [b_\mu c]] = [[a_\lambda b]_{\lambda+\mu} c] + [b_\mu [a_\lambda c]].$ (13.25)

Proof. The first sesquilinearity identity follows from Proposition 13.4(a):

$$[\partial a_\lambda b] = \mathfrak{F}_{z,w}^\lambda [\partial_z a(z), b(w)] = \mathfrak{F}_{z,w}^\lambda \partial_z [a(z), b(w)]$$
$$= -\lambda \mathfrak{F}_{z,w}^\lambda [a(z), b(w)] = -\lambda [a_\lambda b].$$

For the second equality note that

$$[a(z), \partial_w b(w)] = \partial_w [a(z), b(w)],$$

and by Proposition 13.4(a),

$$[a_\lambda \partial b] = \mathfrak{F}_{z,w}^\lambda \partial_w [a(z), b(w)]$$
$$= \lambda \mathfrak{F}_{z,w}^\lambda [a(z), b(w)] + \partial_w \mathfrak{F}_{z,w}^\lambda [a(z), b(w)]$$
$$= (\lambda + \partial) [a_\lambda b].$$

Skewcommutativity follows from Proposition 13.4 (b) by applying $\mathfrak{F}_{z,w}^\lambda$ to both sides of the identity $[b(w), a(z)] = -[a(z), b(w)]$.

Using the Jacobi identity for the commutator

$$[a(z), [b(w), c(u)]] = [[a(z), b(w)], c(u)] + [b(w), [a(z), c(u)]],$$

and Proposition 13.4(c), we get

$$[a_\lambda [b_\mu c]] = \mathfrak{F}_{z,u}^\lambda [a(z), \mathfrak{F}_{w,u}^\mu [b(w), c(u)]]$$
$$= \mathfrak{F}_{z,u}^\lambda \mathfrak{F}_{w,u}^\mu [a(z), [b(w), c(u)]]$$
$$= \mathfrak{F}_{z,u}^\lambda \mathfrak{F}_{w,u}^\mu ([[a(z), b(w)], c(u)] + [b(w), [a(z), c(u)]])$$
$$= \mathfrak{F}_{w,u}^{\lambda+\mu} \mathfrak{F}_{z,w}^\lambda [[a(z), b(w)], c(u)] + \mathfrak{F}_{w,u}^\mu [b(w), \mathfrak{F}_{z,u}^\lambda [a(z), c(u)]]$$
$$= [[a_\lambda b]_{\lambda+\mu} c] + [b_\mu [a_\lambda c]]. \qquad \blacksquare$$

Remark 13.3. One can define the λ-bracket of any pair (a, b) of $\mathfrak{g}$-valued formal distributions as a formal power series in λ, given by the right-hand side of formula (13.21). Then the sesquilinearity and the Jacobi identity still hold, however we need locality in order to prove the skewcommutativiy of the λ-bracket. $\blacksquare$

Definition 13.7. A *Lie conformal algebra* is a $\mathbb{C}[\partial]$-module $\mathfrak{R}$ with a λ-bracket

$$\mathfrak{R} \otimes \mathfrak{R} \longrightarrow \mathbb{C}[\lambda] \otimes \mathfrak{R}$$
$$a \otimes b \longmapsto [a_\lambda b],$$

such that the sesquilinearity, the skewcommutativity and the Jacobi identity hold for all elements of $\mathfrak{R}$.

We now consider some examples of Lie conformal algebras.

Example 13.1. Let $\{d_n \, (n \in \mathbb{Z}), C\}$ be the basis of the Virasoro algebra, defined in §1.3. Let

$$L(z) = \sum_{n \in \mathbb{Z}} d_n z^{-n-2}. \tag{13.26}$$

This formal distribution is called the *Virasoro formal distribution*.

It is straightforward to check that relations (1.21) are equivalent to

$$[L(z), L(w)] = \partial_w L(w) \, \delta(z, w) + 2 \, L(w) \, \partial_w \delta(z, w) + \frac{C}{12} \, \partial_w^3 \delta(z, w). \tag{13.27}$$

We do it as follows. In (13.27) the coefficients of the OPE for the pair (L, L) are

$$L_{(0)}L = \partial L, \quad L_{(1)}L = 2L, \quad L_{(3)}L = \frac{C}{2}, \quad L_{(j)}L = 0$$

$$\text{for } j = 2 \text{ and for } j \geq 4. \tag{13.28}$$

Using the rules (13.17) and (13.18), write the corresponding commutator formula

$$[L_{(m)}, L_{(n)}] = \binom{m}{0}(\partial L)_{(m+n)} + \binom{m}{1} 2L_{(m+n-1)} + \binom{m}{3} \frac{C}{2} \delta_{m+n-3,-1}$$

$$= -(m+n)L_{(m+n-1)} + 2mL_{(m+n-1)}$$

$$+ \frac{(m-1)^3 - (m-1)}{12} C \, \delta_{m-1,-n+1}.$$

Noting that $L_{(m)} = d_{m-1}$ we obtain the equivalence of (13.27) and (1.21).

The OPE (13.27) implies that the pair (L, L) is local. The λ-bracket for $L(z)$ is

$$[L_\lambda L] = (\partial + 2\lambda) L + \frac{\lambda^3}{12} C. \tag{13.29}$$

The Virasoro formal distribution $L(z)$ along with its derivatives and together with the central element C form a Lie conformal algebra

$$Vir = \mathbb{C}[\partial]L \oplus \mathbb{C}C,$$

where $\partial C = 0$, and the λ-bracket is defined by (13.29) and by $[Vir_\lambda C] = 0$.

Example 13.2. Let $\mathfrak{g}$ be a finite dimensional Lie algebra with a non-degenerate symmetric invariant bilinear form $(\cdot|\cdot)$. The *Kac-Moody affinization* of the pair $(\mathfrak{g}, (\cdot|\cdot))$ was introduced in §10.1. It is the central extension $\widehat{\mathfrak{g}}' = \widetilde{\mathfrak{g}} \oplus \mathbb{C}M$ of the loop algebra $\widetilde{\mathfrak{g}} = \mathfrak{g}[t, t^{-1}]$ by a one-dimensional center $\mathbb{C}M$, the commutation relations being given by relations (10.1), if we let $x(k) = xt^k$ for $x \in \mathfrak{g}$ and $k \in \mathbb{Z}$.

Currents are $\widehat{\mathfrak{g}}'$-valued formal distributions of the form

$$a(z) = \sum_{n \in \mathbb{Z}} (at^n)z^{-n-1}, \qquad \text{where } a \in \mathfrak{g}. \tag{13.30}$$

Again, using the rules (13.17) and (13.18), it is immediate to see that relations (10.1) are equivalent to the following OPE:

$$[a(z), b(w)] = [a, b](w)\delta(z, w) + (a|b)\partial_w\delta(z, w)M.$$

This OPE is encoded by the following λ-bracket:

$$[a_\lambda b] = [a, b] + (a|b)\lambda M, \tag{13.31}$$

where $[a, b]$ is the Lie bracket in $\mathfrak{g}$. The local family of currents on $\mathfrak{g}$ along with their derivatives and together with the central element M form a Lie conformal algebra

$$\text{Cur } \mathfrak{g} = \mathbb{C}[\partial]\,\mathfrak{g} \oplus \mathbb{C}M,$$

where $\partial M = 0$, and the λ-bracket is defined by (13.31) and by $[\text{Cur}\mathfrak{g}_\lambda M] = 0$.

When $\dim \mathfrak{g} = 1$, the Kac-Moody affinization is the oscillator algebra $\mathscr{A}$ (cf. Lecture 2). In this case we get the Lie conformal algebra of a *free boson* with the λ-bracket

$$[a_\lambda b] = (a|b)\lambda \mathbb{1}, \ a, b \in \mathfrak{g},$$

where, as in the case of the oscillator algebra, we denote the central element by $\mathbb{1}$.

In order to introduce the third example, that of free fermions, we need a digression to superalgebras.

A vector space V, decomposed in a direct sum of two subspaces:

$$V = V_{\bar{0}} \oplus V_{\bar{1}},$$

is called a *vector superspace*. Here $\bar{0}$ and $\bar{1}$ are elements of the field $\mathbb{Z}/2\mathbb{Z}$, and for $v \in V_\alpha$, $\alpha \in \mathbb{Z}/2\mathbb{Z}$, one calls α the *parity* of v and denotes it by $p(v)$. A *superalgebra* is a superspace $A = A_{\bar{0}} \oplus A_{\bar{1}}$ with a product, such that $A_\alpha A_\beta \subset A_{\alpha+\beta}$. Given a superalgebra, one endows it with a bracket

$$[a, b] = ab - (-1)^{p(a)p(b)} ba. \tag{13.32}$$

This is the usual Lie algebra bracket if $A = A_{\bar{0}}$ is purely even, but in general it is different. Formula (13.32) defines the brackets for homogeneous elements, i.e. when $a \in A_\alpha$, $b \in A_\beta$, and extends to arbitrary elements by bilinearity. Due to its importance, one introduces a notation for the sign in (13.32):

$$p(a, b) = (-1)^{p(a)p(b)}. \tag{13.33}$$

One checks directly that for an associative superalgebra A the bracket (13.32) satisfies the axioms of a Lie superalgebra:

(skewcommutativity) $[b, a] = -p(a, b)[a, b]$,

(Jacobi identity) $[a, [b, c]] = [[a, b]c] + p(a, b)[b, [a, c]]$.

In this example we can observe the general sign rule: one converts an identity for an algebra to one for a superalgebra by changing sign when the order of two odd elements is reversed.

The basic example of an associative superalgebra is $\operatorname{End} V$, where $V = V_{\bar{0}} \oplus V_{\bar{1}}$ is a vector superspace: one lets $(\operatorname{End} V)_\alpha$ consist of all endomorphisms a, such that $a V_\beta \subset V_{\alpha+\beta}$, $\alpha, \beta \in \mathbb{Z}/2\mathbb{Z}$.

A *representation* of an associative superalgebra U (resp. Lie superalgebra $\mathfrak{g}$) in a vector superspace V is a homomorphism of U (resp. $\mathfrak{g}$) to the associative superalgebra $\operatorname{End} V$ (resp. Lie superalgebra $\operatorname{End} V$, endowed with the bracket (13.32)).

If U is a vector superspace, one always assumes that all coefficients of a formal U-valued distribution $a(z) = \sum_n a_{(n)} z^{-n-1}$ have the same parity, which one denotes by $p(a)$. Obviously $\partial_z a(z)$ and $a(z)$ have the same parity.

All statements of this lecture hold in the more general situation when $\mathfrak{g}$ is a Lie superalgebra, if the sign rule is applied. For example, in Proposition 13.5 one should insert the sign $p(a,b)$ in the right-hand side of the skewcommutativity and in the second term of the Jacobi identity:

$$[b_\lambda a] = -p(a,b) [a_{-\lambda-\partial} b] \quad \text{if } a, b \text{ is a local pair,} \qquad (13.34)$$

$$[a_\lambda [b_\mu c]] = [[a_\lambda b]_{\lambda+\mu} c] + p(a,b) [b_\mu [a_\lambda c]]. \qquad (13.35)$$

Consequently, a Lie conformal superalgebra is defined as a $\mathbb{Z}/2\mathbb{Z}$-graded $\mathbb{C}[\partial]$-module, endowed with a λ-bracket, satisfying the axioms of a Lie conformal algebra with modified signs.

We shall often drop the prefix "super" if no confusion may arise.

Example 13.3. Let $A = A_{\bar{0}} \oplus A_{\bar{1}}$ be a vector superspace with a non-degenerate skew-supersymmetric bilinear form, i.e. $\langle . \, | \, . \rangle$:

$$\langle a \, | \, b \rangle = -p(a,b) \langle b \, | \, a \rangle \quad (\text{hence} \quad \langle A_{\bar{0}} \, | \, A_{\bar{1}} \rangle = 0).$$

The *Clifford affinization* C_A of A is a Lie superalgebra

$$C_A = (A[t, t^{-1}]) \oplus \mathbb{C} \cdot \mathbb{1}$$

with parity $p(t^m a) = p(a)$, $p(\mathbb{1}) = \bar{0}$, and the following commutation relations:

$$[at^m, bt^n] = \delta_{m,-n-1} \langle a | b \rangle \mathbb{1}, \quad [at^m, \mathbb{1}] = 0, \ a, b \in A, \ m, n \in \mathbb{Z}. \qquad (13.36)$$

As in the case of bosons, we use $\mathbb{1}$ for the central element since in all representations of C_A that we shall consider $\mathbb{1}$ will be represented by the identity operator.

The form $\langle . \, | \, . \rangle$ is skew-supersymmetric, hence this bracket is skew-commutative, and the Jacobi identity trivially holds since all triple brackets are zero. The formal distributions of the form

$$a(z) = \sum_{n \in \mathbb{Z}} (at^n) z^{-n-1} \qquad (a \in A) \qquad (13.37)$$

are called *free fermions*. The OPE is given by

$$[a(z), b(w)] = \langle a|b\rangle\delta(z,w)\mathbb{1},$$

hence the λ-bracket is

$$[a_{\,\lambda}\, b] = \langle a|b\rangle\mathbb{1}, \quad \text{where } a, b \in A. \tag{13.38}$$

This local family of formal distributions together with the central element $\mathbb{1}$ form a Lie conformal superalgebra of free fermions

$$F(A) = (\mathbb{C}[\partial]\, A) \oplus \mathbb{C}\mathbb{1},$$

where $\partial\mathbb{1} = 0$ and the parity is defined by

$$p(\mathbb{1}) = 0 \quad \text{and} \quad p(f(\partial)a) = p(a), \quad \text{for } a \in A,\; f(\partial) \in \mathbb{C}[\partial].$$

LECTURE 14

14.1. Completion of U, restricted representations and quantum fields

In this lecture we introduce the notion of normal ordered product of continuous formal distributions. This operation simplifies considerably computations with normal orderings in the first part of the book.

Normal ordered product can be defined in the following general setup. Let U be a unital associative superalgebra with a collection of subalgebras $\{U_i\}_{i \in I}$, satisfying the following two properties:

$$\bigcap_{i \in I} U_i = 0, \tag{14.1}$$

and

for any $u \in U$ and any U_k there exists U_l such that $U_l u \subset U U_k$. (14.2)

The main examples of such algebras will come as superalgebras of endomorphisms of vector superspaces and as universal enveloping algebras of $\mathbb{Z}$-graded Lie superalgebras.

Let $\mathrm{End}\, V$ be the superalgebra of endomorphisms of a vector superspace V. For any finite-dimensional subspace W of V denote by $(\mathrm{End}\, V)_W$ the set of all endomorphisms that annihilate the elements of W. The algebra $U = \mathrm{End}\, V$ with the collection of subalgebras $\{U_W = (\mathrm{End}\, V)_W\}$, labeled by the set of all finite-dimensional subspaces of V is an example of an algebra that satisfies properties (14.1) and (14.2). Indeed, the intersection of all subalgebras $(\mathrm{End}\, V)_W$ over all finite-dimensional vector spaces W is obviously zero. For any $a \in \mathrm{End}\, V$ and any finite-dimensional subspace W

of V consider $W' = a(W)$. The subspace W' is finite-dimensional and

$$(\operatorname{End} V)_{W'} a \subset (\operatorname{End} V)_W.$$

This implies property (14.2).

Now we turn to the universal enveloping superalgebras of Lie superalgebras. Recall that, choosing an ordered basis $v_1, v_2, v_3, \ldots$ of a Lie superalgebra $\mathfrak{g}$, compatible with parity, the Poincaré-Birkhoff-Witt (PBW) theorem states that the corresponding basis of the universal enveloping superalgebra $U(\mathfrak{g})$ consists of ordered monomials

$$v_1^{n_1} v_2^{n_2} v_3^{n_3} \ldots, \quad \text{where } n_i \in \mathbb{Z}_+ \text{ and } n_i \leq 1 \quad \text{if} \quad p(v_i) = \bar{1}. \tag{14.3}$$

Let $\mathfrak{a}$ be a $\mathbb{Z}$-graded Lie superalgebra:

$$\mathfrak{a} = \bigoplus_{j \in \mathbb{Z}} \mathfrak{a}_j, \quad [\mathfrak{a}_i, \mathfrak{a}_j] \subset \mathfrak{a}_{i+j}. \tag{14.4}$$

Consider the universal enveloping superalgebra $U = U(\mathfrak{a})$ with the collection of subalgebras $\{U_l\}_{l=1,2,\ldots}$ defined as universal enveloping superalgebras of the form

$$U_l := U\left(\bigoplus_{j \geq l} \mathfrak{a}_j\right), \quad l = 1, 2 \ldots. \tag{14.5}$$

These subalgebras have zero intersection. Indeed, choose a basis $\{v_s\}$ of the Lie superalgebra $\mathfrak{a}$, compatible with the parity, with the property that in the grading (14.4), $\deg v_s \leq \deg v_t$ whenever $s < t$. Suppose that there exists a nonzero element u in the intersection $\bigcap_{i \in I} U_i$. By the PBW theorem u can be written uniquely as a linear combination of ordered monomials (14.3). Let v_s be the highest degree element in the grading (14.4) that appears in this linear combination, and let $p = \deg v_s$. Then $u \notin U_{p+1}$, and we get a contradiction.

Property (14.2) holds for the subalgebras (14.5) as well, since if $u \in U$ is a product of n homogeneous elements of $\mathfrak{a}$, each of degree greater or equal to $-M$ with $M \geq 0$, we have by induction on n: $U_l u \subset U U_{l-nM}$.

The Virasoro algebra Vir, the Kac-Moody affinization $\hat{\mathfrak{g}}'$ and the Clifford affinization C_A are Lie superalgebras with a natural $\mathbb{Z}$-grading. Namely, the grading components are, respectively, for $\mathfrak{a} = Vir$:

$$\mathfrak{a}_0 = \mathbb{C}C \oplus \mathbb{C}d_0, \quad \text{and } \mathfrak{a}_i = \mathbb{C}d_i \quad \text{for} \quad i = \pm 1, \pm 2, \ldots,$$

for the Kac-Moody affinization $\mathfrak{a} = \hat{\mathfrak{g}}' = \mathfrak{g}[t, t^{-1}] \oplus \mathbb{C}M$ of a Lie algebra $\mathfrak{g}$:

$$\mathfrak{a}_0 = \mathbb{C}M \oplus \mathfrak{g}, \quad \text{and} \quad \mathfrak{a}_i = \mathfrak{g}t^i, \quad \text{for} \quad i = \pm 1, \pm 2, \ldots,$$

and for the Clifford affinization $\mathfrak{a} = C_A = A[t, t^{-1}] \oplus \mathbb{C} \cdot \mathbb{1}$ of a vector superspace A:

$$\mathfrak{a}_0 = \mathbb{C} \cdot \mathbb{1}, \quad \mathfrak{a}_{2i+1} = At^i, \quad \text{and} \quad \mathfrak{a}_{2i} = 0 \quad \text{for} \quad i = \pm 1, \pm 2, \pm 3, \ldots.$$

Let U be a superalgebra with a collection of subalgebras satisfying properties (14.1) and (14.2).

Definition 14.1. The *completion* of U is the space U^c of infinite series $\sum_i u_i$ such that for each U_k, $k \in I$, all but finitely many terms u_i lie in UU_k. Two series $\sum_i u_i$ and $\sum_i v_i$ are said to be equal in U^c, if their images (which are finite sums) in any of U/U_j $(j \in I)$ are equal.

The space U^c is an algebra with an obvious product. Indeed, let $\sum_i u_i$ and $\sum_j v_j$ be two elements of U^c. Fix $k \in I$. By assumption, all but finitely many elements v_j are in UU_k. For those v_j that lie in UU_k and for any u_i, one has $u_i v_j \in U U_k$. Let $v_{j_1}, \ldots, v_{j_m}$ be all the terms that are not in $U U_k$. By property (14.2), there exist $l_1, \ldots, l_m \in I$ such that $U_{l_s} v_{j_s} \subset U U_k$. For each of $s = 1, \ldots, m$ all but finitely many u_i's lie in $U U_{l_s}$ and, hence, $u_i v_{j_s} \in U U_k$ for all but finitely many u_i's. Thus $\sum_i u_i \sum_j v_j$ also lies in the completion U^c.

Due to condition (14.1), U is an embedded subalgebra of U^c.

Note that the completion $(\text{End}\, V)^c$ coincides with the algebra $\text{End}\, V$. In general, we will be interested in representations of the algebra U in a vector space V, which extend to its completion U^c. The following is an important class of such representations.

Definition 14.2. A representation of the algebra U in a vector space V is called *restricted* if for any $v \in V$ there exists $i \in I$, such that $U_i v = 0$.

The reader can easily check that all representations of the oscillator algebra, the affine Lie algebras and the Virasoro algebra, considered in Lectures 1-12, give rise to restricted representations of their universal enveloping algebras, with the exception of representations $V_{\alpha,\beta}$ of the Virasoro algebra in §1.2. In particular, admissible representations of affine Lie algebras $\hat{\mathfrak{g}}$, as defined in Lecture 10, give rise to restricted representations of $U(\hat{\mathfrak{g}})$ and vice versa.

It is clear that any subrepresentation and factor representation of a restricted representation are restricted.

Remark 14.1. Taking the collection of subalgebras $\{U_i\}_{i \in I}$ for a fundamental system of neighborhoods of 0 makes U a topological algebra. A restricted representation of U in a vector space V is just a continuous representation, if we endow V with discrete topology. ∎

Definition 14.3. A U^c-valued formal distribution $a(z) = \sum_n a_{(n)} z^{-n-1}$ is called *continuous* if for any U_k, $a_{(n)} \in U^c U_k$ for n sufficiently large.

The Virasoro formal distribution (13.26), the currents (13.30) and the free fermions (13.37) are examples of continuous formal distributions.

Definition 14.4. Let V be a vector superspace. A *quantum field* is a continuous formal $\operatorname{End} V$-valued distribution. Equivalently, it is a formal $\operatorname{End} V$-valued distribution $a(z) = \sum_{n \in \mathbb{Z}} a_{(n)} z^{-n-1}$ such that for any vector $v \in V$, $a_{(n)} v = 0$ for n sufficiently large, i.e. for any $v \in V$, the formal V-valued distribution $a(z)v$ is a formal Laurent series in z.

It is clear from the definitions that in a restricted representation of U in a vector superspace V the image of a continuous U^c-valued formal distribution is a quantum field, and the image of a local pair of continuous formal distributions is a local pair of quantum fields. This observation leads to one of the most important constructions of quantum fields and local pairs of quantum fields.

Recall that, given an associative superalgebra U, its subalgebra W, and a representation of the algebra W in a vector superspace V_0, one defines the *induced representation* $\operatorname{Ind}_W^U V_0$ of U as the factor-space of the space $U \otimes V_0$ by the subspace spanned by elements $uw \otimes v_0 - u \otimes wv_0$ with $u \in U$, $w \in W$, $v_0 \in V_0$, and the action of U is defined via left multiplication in U. Note that if $\{u_i\}$ is a basis of $U \bmod W$ and $\{v_j\}$ is a basis of V_0, then $\{u_i(v_j)\}$ is a basis of $\operatorname{Ind}_W^U V_0$.

Given a Lie superalgebra $\mathfrak{g}$, its subalgebra $\mathfrak{h}$, and a representation of $\mathfrak{h}$ in V_0, we have the universal enveloping superalgebra $U = U(\mathfrak{g})$ and a representation of its subalgebra $W = U(\mathfrak{h})$ in V_0. In this case $\operatorname{Ind}_W^U V_0$ is usually denoted by $\operatorname{Ind}_{\mathfrak{h}}^{\mathfrak{g}} V_0$.

Proposition 14.1. Assume that the collection of subalgebras $\{U_i\}_{i \in I}$ of the associative superalgebra U is nested, i.e. it is of the form

$$U_1 \supset U_2 \supset U_3 \supset \ldots.$$

Let W be a subalgebra of U, containing U_j for some $j \in I$. Consider a restricted representation of W in a vector space V_0. Then the induced representation $V = \operatorname{Ind}_W^U V_0$ of the algebra U is restricted. Consequently, the representation of U in V extends to U^c, and continuous U^c-valued formal distributions map to quantum fields in this representation.

Proof. We need to prove that a vector $v = u_1 \ldots u_k v_0 \in V$, where $u_1, \ldots, u_k \in U$, and $v_0 \in V_0$, is annihilated by U_l for some l. We prove this claim by induction on k. For $k = 0$, $v = v_0$, hence, the claim holds since V_0 is a restricted representation of W. If $k \geq 1$, by the inductive assumption, the claim holds for the vector $v_1 = u_2 \ldots u_k v_0$, i.e. $U_k v_1 = 0$ for some k. But then by (14.2), $U_l u_1 \subset U U_k$ for some l. Hence $U_l v = U_l u_1 v_1 \subset U U_k v_1 = 0$.

∎

Corollary 14.1. Let U be an associative superalgebra with a nested collection of subalgebras $\{U_i\}_{i \in I}$, and let V be a representation of U, for which there exists a vector v, such that $Uv = V$ and $U_j v = 0$ for some $j \in I$. Then the representation V is restricted.

Proof. V is a factor representation of the representation of U, induced from the trivial one-dimensional representation of U_j. ∎

Example 14.1. The Virasoro algebra (see Example 13.1) contains a subalgebra, often denoted by Vir_-, which is spanned by the elements d_n with $n \geq -1$. Given $c \in \mathbb{C}$, consider the one-dimensional representation $\mathbb{C}_c$ of the subalgebra $Vir_- \oplus \mathbb{C}C$, defined by

$$d_n = 0, \quad C = c.$$

Then the induced representation

$$V^c = \operatorname{Ind}_{Vir_- \oplus \mathbb{C}C}^{Vir} \mathbb{C}_c$$

is a restricted representation, called *the vacuum representation* of the Virasoro algebra with *central charge c*. The representation V^c is not irreducible if and only if $c = c(m)$, given by (8.17), where $m + 2$ is a nonnegative rational number of the form $\frac{q}{p-q}$, where $p > q > 1$, p and q are coprime integers. In other words, V^c is not irreducible if and only if

$$c = 1 - \frac{6(p-q)^2}{pq}, \text{ where } p \text{ and } q \text{ are coprime integers} \geq 2 \qquad (14.6)$$

(Gorelik-Kac [2007]). These coincide with the minimal series central charges of Belavin-Polyakov-Zamolodchikov [1984a]. Recall that it is shown in Lecture 12 that these representations are unitary if m is a nonnegative integer.

Example 14.2. The Kac-Moody affinization $\hat{\mathfrak{g}}'$ (see Example 13.2) contains a subalgebra $\mathfrak{g}[t]$. Given $m \in \mathbb{C}$, consider the one-dimensional representation $\mathbb{C}_m$ of the subalgebra $\mathfrak{g}[t] \oplus \mathbb{C}M$, defined by

$$\mathfrak{g}[t] = 0, \quad M = m.$$

Then the induced representation

$$V^m(\mathfrak{g}) = \operatorname{Ind}_{\mathfrak{g}[t] \oplus \mathbb{C}M}^{\hat{\mathfrak{g}}'} \mathbb{C}_m$$

is a restricted representation, called *the vacuum representation* of $\hat{\mathfrak{g}}'$ of *level m*. One can find the conditions of irreducibility of representations $V^m(\mathfrak{g})$ in Gorelik-Kac [2007] as well, provided that $\mathfrak{g}$ is a simple Lie algebra. For example, if $\mathfrak{g} = sl_n$ with the invariant bilinear form (9.18), then the representation $V^m(\mathfrak{g})$ is not irreducible if and only if $m + n$ is not a nonnegative rational number, different from a reciprocal of a positive integer.

Example 14.3. The free boson (see Example 13.2) is a slight generalization of a particular case of the previous example, when $\mathfrak{g}$ is one-dimensional, $\mathfrak{g} = \mathbb{C}a$, $(a|a) = 1$. Then the elements $a_n = a \otimes t^n$ $(n \in \mathbb{Z})$ and $M = \mathbb{1}$ form a standard basis of the oscillator algebra $\mathscr{A}$. Given $\mu \in \mathbb{C}$, consider the one-dimensional representation $\mathbb{C}(\mu)$ of the (abelian) subalgebra $\mathbb{C}\mathbb{1} + \sum_{n \geq 0} \mathbb{C}a_n$, defined by

$$a_n = 0 \quad \text{for} \quad n > 0, \quad a_0 = \mu, \quad \mathbb{1} = 1.$$

Then the induced representation

$$B^{(\mu)} = \operatorname{Ind}_{\mathbb{C}\mathbb{1} + \sum_{n \geq 0} \mathbb{C}a_n}^{\mathscr{A}} \mathbb{C}(\mu)$$

is equivalent to the representation of $\mathscr{A}$, defined by formulas (2.4 a-c). This follows from Proposition 2.2, since vectors $a_{-n_1} \ldots a_{-n_k}(1)$ with $1 \leq$

$n_1 \leq \ldots \leq n_k$ form a basis of $B^{(\mu)}$. This representation is restricted and irreducible for all μ, and it is called a *bosonic Fock representation of charge* μ.

Example 14.4. Let A be a vector superspace endowed with a non-degenerate skew-supersymmetric bilinear form $\langle \cdot | \cdot \rangle$, and let C_A be its Clifford affinization, considered in Example 13.3. Consider the one-dimensional representation $\mathbb{C}$ of the (abelian) subalgebra $A[t] \oplus \mathbb{C}\mathbb{1}$, defined by:

$$A[t] = 0, \quad \mathbb{1} = 1.$$

Then the induced representation

$$F(A) = \operatorname{Ind}_{A[t] \oplus \mathbb{C}\mathbb{1}}^{C_A} \mathbb{C}$$

is a restricted representation of the Lie superalgebra C_A. This representation is irreducible. The proof of this fact is similar to that in the previous example (see also proof of Theorem 16.2 in §16.6). Note that only in this example the induced representation is not a purely even vector superspace, the parity on $F(A)$ being induced by that on A by letting $p(\mathbb{C}) = \bar{0}$. This representation is called a *fermionic Fock representation*.

In all these examples the basic formal distributions, namely the Virasoro distribution, currents, free boson, and free fermions, are continuous and their images in the corresponding induced representations are pairwise local quantum fields. These quantum fields are called respectively Virasoro quantum field, currents, free boson of charge μ, and free fermions.

14.2. Normal ordered product

Let $a(z)$ be a formal distribution with coefficients in a vector superspace:

$$a(z) = \sum_{n \in \mathbb{Z}} a_{(n)} z^{-n-1}.$$

We split this formal distribution into two parts:

$$a(z)_+ = \sum_{n \leq -1} a_{(n)} z^{-n-1}, \qquad a(z)_- = \sum_{n \geq 0} a_{(n)} z^{-n-1}. \tag{14.7}$$

Note that this is the only way to split the formal distribution into two parts that are respected by the operation of differentiation:

$$\partial_z \left(a(z)_\pm\right) = \left(\partial_z\, a(z)\right)_\pm. \tag{14.8}$$

Moreover, the series $a(z)_\pm$ satisfy the properties

$$\text{Res}_z\, a(z)\, \iota_{z,w}\, \frac{1}{(z-w)^{n+1}} = \frac{\partial_w^n}{n!}\, a(w)_+, \tag{14.9}$$

$$\text{Res}_z\, a(z)\, \iota_{w,z}\, \frac{1}{(z-w)^{n+1}} = -\frac{\partial_w^n}{n!}\, a(w)_-. \tag{14.10}$$

These are called the *Formal Cauchy formulas*, since they are similar to Cauchy formulas in complex analysis.

In order to prove the first formula for $n = 0$ substitute in (14.9) the expansion (13.7). The left-hand side becomes

$$\text{Res}_z\, a(z)\frac{1}{z}\sum_{j\geq 0}\left(\frac{w}{z}\right)^j = \text{Res}_z\, \frac{1}{z}\sum_{m\in\mathbb{Z},j\geq 0} a_{(m)}w^j z^{-j-m-1}$$

$$= \sum_{m\leq -1} a_{(m)}w^{-m-1} = a(w)_+ .$$

The case of $n > 0$ follows by successive differentiation. A similar proof holds for the second Formal Cauchy formula.

Now we consider formal distributions with values in the completion U^c of the associative superalgebra U.

Definition 14.5. The *normal ordered product* of U^c-valued formal distributions $a(z)$ and $b(z)$ is defined by

$$: a(z)b(z) : = a(z)_+b(z) + p(a,b)b(z)a(z)_-. \tag{14.11}$$

Example 14.5. Let $\mathscr{A}$ be the oscillator algebra defined in §2.2. The basis $\mathbb{1}$ and $\{a_n\}_{n\in\mathbb{Z}}$ of $\mathscr{A}$ satisfy relations (2.2). We set $\mathbb{1} = 1$, by which we mean that we work in the quotient U of the universal enveloping algebra $U(\mathscr{A})$ by the ideal generated by $(\mathbb{1}-1)$, or rather in its completion U^c. As has been mentioned in §10.1, the oscillator algebra can be viewed as the Kac-Moody affinization of a one-dimensional abelian Lie algebra $\mathbb{C}a$ with the bilinear form $(a\,|\,a) = 1$. Here $a_n = a\otimes t^n$. The current (free boson)

$\alpha(z) = \sum_{n \in \mathbb{Z}} a_n z^{-n-1}$ satisfies

$$[\alpha(z), \alpha(w)] \quad = \quad \partial_w \, \delta(z, w) \cdot 1,$$

or, in terms of λ-brackets,

$$[\alpha_\lambda \alpha] \quad = \quad \lambda \cdot 1. \tag{14.12}$$

Put

$$L(z) \quad = \quad \frac{1}{2} : \alpha(z)\alpha(z) :$$

and write $L(z) = \sum_{k \in \mathbb{Z}} L_k z^{-k-2}$. The coefficients L_k of this normal ordered product are exactly the Virasoro operators discussed in §2.3; see formulas (2.9) and (2.10) there.

The normal ordered product of two U^c-valued formal distributions may diverge. However for the continuous formal distributions it is well defined.

Lemma 14.1. (a) The normal ordered product of two continuous formal distributions is a continuous formal distribution.
(b) The derivative ∂_z of a continuous formal distribution is again a continuous formal distribution.

Proof. (a) Let $c(z) = : a(z)b(z) : = \sum_{l \in \mathbb{Z}} c_{(l)} z^{-l-1}$. Then

$$c_{(l)} = \sum_{j \leq -1} a_{(j)} b_{(l-j-1)} + p(a, b) \sum_{j \geq 0} b_{(l-j-1)} a_{(j)}. \tag{14.13}$$

First, let us show that $c(z)$ is a U^c-valued formal distribution. Since $a(u)$ and $b(u)$ are continuous formal distributions, the terms $a_{(j)}$'s for big enough j and the terms $b_{(l-j-1)}$'s for big enough $(l-j-1)$ lie in $U^c U_k$. Hence, in (14.13) all but finitely many terms of both sums lie in $U^c U_k \subset U^c$. Let the finite sum

$$\sum_{-M \leq j \leq -1} a_{(j)} b_{(l-j-1)} + p(a, b) \sum_{M \geq j \geq 0} b_{(l-j-1)} a_{(j)}$$

include all the terms that are left. Observe that this is a linear combination of products of elements of U^c of the form $a_{(j)} b_{(l-j-1)}$ and $b_{(l-j-1)} a_{(j)}$. Since U^c is an algebra, this linear combination is also an element of U^c. Hence, $c(z)$ is a U^c-valued formal distribution.

Let us prove that $c(z)$ is continuous. Since $b(z)$ is a continuous formal distribution, there exists a positive integer N, such that for any $m \geq N$, $b_{(m)} \in U^c U_k$. Then for any $l \geq N$ the sum of terms over $j \leq -1$ in (14.13) is in $U^c U_k$.

By the continuity of $a(z)$, there exists M, such that in the sum (14.13) over $j \geq 0$, for any $j > M$ the terms $b_{(l-j-1)}a_{(j)}$ lie in $U^c U_k$. The elements $a_{(0)}, \ldots, a_{(M)}$ are elements of U^c, and by definition, all but finitely many terms of these series lie in $U U_k$. Let $\{a_{(j)}^s\}$ be this finite set that is left. By property (14.2), for any such $a_{(j)}^s$ there exists $U_{l_{j,s}}$ such that $U_{l_{j,s}} a_{(j)}^s \in U U_k$.

By the continuity of $b(z)$, there exist positive integers $K_{j,s}$, such that for all $m \geq K_{j,s}$, $b_{(m)} \in U^c U_{l_{j,s}}$, and $b_{(m)} a_{(j)}^s \in U^c U_k$. Then if $l \geq M + \max_{j,s}\{K_{j,s}\}$, any element in the sum over $j \geq 0$ of (14.13) is in $U^c U_k$.

Summarizing all these arguments, we obtain, that $c_{(l)} \in U^c U_k$ for all $l \geq \max\{N, M + K_{j,s}\}$, and the formal distribution is continuous. (b) is obvious. ∎

The normal ordered product of formal distributions allows to extend Definition 13.3 of the n-th product to the negative values of n. Namely, for $n \in \mathbb{Z}_+$ we let

$$a(w)_{(-n-1)}b(w) =: \frac{\partial_w^n a(w)}{n!} b(w) : .$$

Proposition 14.2. Let $a(z)$ and $b(z)$ be U^c-valued continuous formal distributions. Then
(a) For any $n \in \mathbb{Z}$ the *n-th product* is given by the formula

$$a(w)_{(n)}b(w) = \mathrm{Res}_z \left(a(z)b(w)\iota_{z,w}(z-w)^n \right.$$
$$\left. - p(a,b)b(w)a(z)\iota_{w,z}(z-w)^n \right). \tag{14.14}$$

(b) The n-th product $a(w)_{(n)}b(w)$ is a U^c-valued continuous formal distribution.
(c) The operator ∂_w is a derivation of the n-th product for any $n \in \mathbb{Z}$:

$$\partial_w(a(w)_{(n)}b(w)) = (\partial_w a(w))_{(n)}b(w) + a(w)_{(n)}(\partial_w b(w)).$$

Proof. Statement (a) for $n \geq 0$ is obvious, and for $n \leq -1$ follows from Formal Cauchy formulas (14.9), (14.10). Then the continuity property (b) for $n \geq 0$ follows from the definition, and for $n \leq -1$ from the continuity of the normal ordered product (Lemma 14.1 (a)). Statement (c) follows from Proposition 13.2 for $n \geq 0$ and from (14.8) for $n \leq -1$. ∎

Corollary 14.2. (B. Bakalov) If a pair of continuous U^c-valued distributions is local, i.e. for any sufficiently large $N \in \mathbb{Z}_+$

$$(z - w)^N a(z) b(w) = p(a, b)(z - w)^N b(w) a(z), \qquad (14.15)$$

then for any $k \in \mathbb{Z}_+$ one has:

$$a(w)_{(N-1-k)} b(w) = \mathrm{Res}_z \, F(z, w) \partial_w^k \delta(z, w)/k!, \qquad (14.16)$$

where $F(z, w)$ denotes either side of the equality (14.15).

Proof. It follows from (14.14) and (13.10). ∎

LECTURE 15

15.1. Non-commutative Wick formula

The central result of this lecture is the so-called non-commutative Wick formula, which relates the λ-bracket with the normally ordered product, as does the usual Leibniz-Wick formula, but with a "quantum correction".

Proposition 15.1. Let $a(z)$, $b(z)$, $c(z)$ be continuous formal distributions with values in U^c. For any $n \in \mathbb{Z}$ one has

$$[a_\lambda (b_{(n)} c)] = \sum_{k \in \mathbb{Z}_+} \frac{\lambda^k}{k!} [a_\lambda b]_{(n+k)} c + p(a,b) \, b_{(n)} [a_\lambda c]. \tag{15.1}$$

Proof. By the definition of the λ-bracket,

$$[a_\lambda (b_{(n)} c)] = \mathrm{Res}_z \, e^{\lambda(z-w)} [a(z) , b(w)_{(n)} c(w)]$$
$$= \mathrm{Res}_z \, \mathrm{Res}_x \, e^{\lambda(z-w)} [a(z), b(x) c(w)] i_{x,w} (x-w)^n$$
$$- p(b,c) \mathrm{Res}_z \, \mathrm{Res}_x \, e^{\lambda(z-w)} [a(z), c(w) b(x)] i_{w,x} (x-w)^n.$$

Denote by A the first term in the right-hand side and by B the second term. Using that $[a, bc] = [a, b]c + p(a,b)b[a,c]$ and $e^{z-w} = e^{z-x} e^{x-w}$, we have

$$A = \mathrm{Res}_z \, \mathrm{Res}_x \, e^{\lambda(z-x)} e^{\lambda(x-w)} [a(z), b(x)] \, c(w) i_{x,w} (x-w)^n$$
$$+ p(a,b) \mathrm{Res}_z \, \mathrm{Res}_x \, e^{\lambda(z-w)} b(x) \, [a(z), c(w)] i_{x,w} (x-w)^n$$
$$= \mathrm{Res}_x \, e^{\lambda(x-w)} [a(x)_\lambda b(x)] \, c(w) i_{x,w} (x-w)^n$$
$$+ p(a,b) \mathrm{Res}_x \, b(x) \, [a(w)_\lambda c(w)] i_{x,w} (x-w)^n,$$

163

and similarly,

$$B = -p(b,c)\text{Res}_x \,[a(w)_\lambda c(w)]\, b(x)i_{w,x}(x-w)^n$$
$$- p(b,c)p(a,c)\text{Res}_x \, c(w)e^{\lambda(x-w)}[a(x)_\lambda b(x)]\, i_{w,x}(x-w)^n.$$

Denote by A_1 and A_2 (resp. B_1 and B_2) the first and the second summand in A (resp. in B). Note that

$$A_2 + B_1 = p(a,b)b_{(n)}[a_\lambda c],$$

and, since $e^{\lambda(x-w)} = \sum_{k\geq 0} \frac{\lambda^k}{k!}(x-w)^k$, we have:

$$A_1 + B_2 = \text{Res}_x \sum_{k\geq 0} \frac{\lambda^k}{k!}\left(i_{x,w}(x-w)^{n+k}[a(x)_\lambda b(x)]c(w) \right.$$
$$\left. - p(b,c)\,p(a,c)c(w)[a(x)_\lambda b(x)]i_{w,x}(x-w)^{n+k} \right)$$
$$= \sum_{k\geq 0} \frac{\lambda^k}{k!}[a_\lambda b]_{(n+k)}c.$$

Hence,

$$[a_\lambda (b_{(n)}c)] = A_1 + A_2 + B_1 + B_2 = \sum_{k\geq 0} \frac{\lambda^k}{k!}[a_\lambda b]_{(n+k)}c + p(a,b)\, b_{(n)}[a_\lambda c].$$

∎

Applying (15.1) to each term $b(w)_{(k)}\, c(w)$ of the generating series

$$[b_\mu c] = \sum_{k\in\mathbb{Z}_+} \frac{\mu^k}{k!}\, b(w)_{(k)}\, c(w),$$

we get again the Jacobi identity of Proposition 13.5 (rather formula (13.35)). If we put $n = -1$, the left-hand side of (15.1) becomes $[a_\lambda : bc :]$, while the right-hand side becomes

$$p(a,b) : b\,[a_\lambda c] : + : [a_\lambda b]\, c : + \sum_{k\geq 1} \frac{\lambda^k}{k!}[a_\lambda b]_{(k-1)}\, c.$$

Rewriting the last term in the form

$$\sum_{k \geq 1} \frac{\lambda^k}{k!} [a_\lambda b]_{(k-1)} c = \int_0^\lambda [[a_\lambda b]_\mu c] \, d\mu,$$

we obtain the *non-commutative Wick formula*:

$$[a_\lambda : bc :] \;\; = \;\; : [a_\lambda b] c : + p(a,b) : b [a_\lambda c] : + \int_0^\lambda [[a_\lambda b]_\mu c] \, d\mu. \quad (15.2)$$

Remark 15.1. Multiplying both sides of (15.1) by w^{-n-1} and summing over $n \in \mathbb{Z}$, we obtain the generating series of the identities (15.1):

$$[a_\lambda b(w) c] = e^{\lambda w} [a_\lambda b](w) c + p(a,b) b(w) [a_\lambda c].$$

Dividing both sides of this identity by w and taking the residue by w, we obtain a more conceptual derivation of the non-commutative Wick formula, using the following obvious identity for an arbitrary formal distribution $\phi(w)$:

$$\frac{e^{\lambda w} \phi(w)}{w} = \frac{\phi(w)}{w} + \int_0^\lambda e^{\mu w} \phi(w) d\mu. \qquad \blacksquare$$

The following proposition is also useful for calculations with n-th products.

Proposition 15.2. For any U^c-valued continuous formal distributions $a(z)$, $b(z)$ and for any $n \in \mathbb{Z}$, the following equalities hold:

$$(\text{Sesquilinearity}) \qquad (\partial a)_{(n)} b = -n \, a_{(n-1)} b; \qquad\qquad (15.3)$$
$$a_{(n)} \partial b = \partial \left(a_{(n)} b \right) + n \, a_{(n-1)} b.$$

$$(\text{Quasicommutativity}) \qquad \text{If } a(z), b(z) \text{ is a local pair, then}$$

$$: ab : - p(a,b) : ba : = \int_{-\partial}^0 [a_\lambda b] \, d\lambda. \quad (15.4)$$

(Here $\int_{-\partial}^0 [a_\lambda b] \, d\lambda = - \sum_{n \geq 0} \frac{(-\partial)^{n+1}}{(n+1)!} \left(a_{(n)} b \right)$ by the usual rules of integration.)

Proof. (*Sesquilinearity*) From the definition of n-th product (14.14),

$$(\partial a)_{(n)}\, b = \mathrm{Res}_z\,(i_{z,w}(z-w)^n \partial_z\, a(z)\, b(w))$$
$$- p(a,b)\mathrm{Res}_z\,(i_{w,z}(z-w)^n b(w)\partial_z\, a(z)).$$

Note that

$$\partial_z\,(\,i_{z,w}(z-w)^n a(z)\,)b(w) = i_{z,w}(z-w)^n \partial_z\, a(z)\, b(w)$$
$$+ \partial_z\,(i_{z,w}(z-w)^n)a(z)b(w),$$

and that the operators ∂_z and $i_{z,w}$ commute. Using (13.2), we get

$$\mathrm{Res}_z\, i_{z,w}(z-w)^n \partial_z a(z)\, b(w) = -n\mathrm{Res}_z\, i_{z,w}(z-w)^{n-1}a(z)\, b(w),$$

and, similarly,

$$-p(a,b)\mathrm{Res}_z\, i_{w,z}(z-w)^n\, b(w)\partial a(z) = np(a,b)\mathrm{Res}_z\, i_{w,z}(z-w)^{n-1}\, b(w)\, a(z).$$

Adding the last two expressions, we get the first equality of (15.3). For the proof of the second equality of (15.3), observe that

$$\partial\big(a_{(n)}b\big) = \partial_w\big(\mathrm{Res}_z\,(i_{z,w}(z-w)^n a(z)\, b(w))$$
$$-p(a,b)\mathrm{Res}_z\,(i_{w,z}(z-w)^n b(w)a(z))$$
$$= \mathrm{Res}_z\,\big(n\, i_{z,w}(z-w)^{n-1}a(z)\, b(w) + p(a,b)n\, i_{w,z}(z-w)^{n-1}b(w)\, a(z)\big)$$
$$+\mathrm{Res}_z\,\big(i_{z,w}(z-w)^n a(z)\,\partial_w b(w) + p(a,b)\, i_{w,z}(z-w)^n \partial_w b(w)\, a(z)\big)$$
$$= \partial a_{(n)}b + a_{(n)}\partial b.$$

(*Quasicommutativity*) Consider a formal distribution in two variables

$$a(z,w) = a(z)b(w)\iota_{z,w}\,\frac{1}{z-w} - p(a,b)b(w)a(z)\iota_{w,z}\,\frac{1}{z-w}.$$

This is a local formal distribution, and

$$:a(w)b(w): \,=\, \mathrm{Res}_z\, a(z,w), \qquad :b(w)a(w): \,=\, p(b,a)\,\mathrm{Res}_z\, a(w,z).$$

Then

$$: ab : - p(a,b) : ba : = \text{Res}_z \; (a(z,w) - a(w,z))$$

$$= \text{Res}_z \; \left(e^{\lambda(z-w)}(a(z,w) - a(w,z)) \right)|_{\lambda=0}$$

$$= \left(\mathfrak{F}_{z,w}^{\lambda} a(z,w) - \mathfrak{F}_{z,w}^{-\lambda-\partial_w} a(z,w) \right)|_{\lambda=0}$$

$$= \text{Res}_z \; (1 - e^{-\partial_w \circ (z-w)}) a(z,w)$$

$$= - \sum_{n \geq 1} \frac{(-\partial_w)^n}{n!} \text{Res}_z \; (z-w)^n a(z,w)$$

$$= - \sum_{n \geq 0} \frac{(-\partial)^{n+1}}{(n+1)!} \left(a_{(n)} b \right). \qquad \blacksquare$$

Lemma 15.1. (Dong's lemma) *If a, b and c are three pairwise local continuous formal distributions, then the pair $(a_{(n)}b, c)$ is also local for any $n \in \mathbb{Z}$.*

Proof. Locality of the pair (a, b) means (14.15) for some $N \in \mathbb{Z}_+$, therefore, by Bakalov's formula (14.16), we need to prove that

$$\text{Res}_z \; (x-w)^M (z-w)^N a(z)b(w)c(x)\partial_w^k \delta(z,w)/k!$$

$$= p(a,c)p(b,c)\text{Res}_z \; (x-w)^M (z-w)^N c(x)a(z)b(w)\partial_w^k \delta(z,w)/k! \quad (15.5)$$

for $M \in \mathbb{Z}_+$ sufficiently large.

We can permute $b(w)$ and $c(x)$ in the left-hand side of (15.5) with a sign $p(b,c)$, for M sufficiently large, since the pair (b,c) is local. Since $(x-w)^M = \sum_{j=0}^{M} \binom{M}{j}(x-z)^{M-j}(z-w)^j$, using property (a) of the formal delta function, given by Proposition 13.1, we can further rewrite the left-hand side of (15.5) as

$$p(b,c)\text{Res}_z \sum_{j=0}^{k} \binom{M}{j}(x-z)^{M-j}(z-w)^{j+N} a(z)c(x)b(w)\partial_w^k \delta(z,w)/k!.$$

Hence, for M sufficiently large, we can permute $a(z)$ and $c(x)$ in this expression with a sign $p(a,c)$, to obtain the right-hand side of (15.5). $\quad \blacksquare$

We have the following rules that will be used extensively in calculations with λ-brackets in all subsequent examples:

- sesquilinearity (13.23) of the λ-bracket;
- skewcommutativity (13.34) of the λ-bracket in presence of locality;
- sesquilinearity (15.3) of n-th products, in particular, ∂ is their derivation;
- the non-commutative Wick formula (15.2);
- quasicommutativity (15.4) of normal ordered product in presence of locality;
- quasiassociativity (17.29) of normal ordered product in presence of locality.

The last property has not been proven yet. It follows from the theory of vertex algebras, developed in Lecture 17. Namely, one has to apply Proposition 17.1, Remark 17.2, and Theorem 17.3.

By Dong's lemma, given a collection of pairwise local continuous formal distributions, we can always apply all of the above rules to normal ordered products of these formal distributions.

15.2. Virasoro formal distribution for free boson

Let $\mathscr{A}$ be the oscillator algebra and let $\alpha(z)$ be the formal distribution defined in Example 14.5 with the λ-bracket (14.12). Set

$$L(z) \;=\; \frac{1}{2} : \alpha(z)\alpha(z) : \, .$$

We can compute λ-brackets of distributions $L(z)$ and $\alpha(z)$ using the rules listed above.

By the non-commutative Wick formula and by (14.12),

$$[\alpha_\lambda L] = \frac{1}{2}[\alpha_\lambda : \alpha\alpha :] = \lambda\alpha \,+\, \frac{1}{2}\int_0^\lambda [[\alpha_\lambda\alpha]_\mu\alpha]\,d\mu. = \lambda\alpha.$$

By skewcommutativity of the λ-bracket,

$$[L_\lambda\alpha] \;=\; -[\alpha_{-\lambda-\partial}L] \;=\; (\lambda+\partial)\alpha. \tag{15.6}$$

Using again the non-commutative Wick formula we get

$$[L_\lambda L] = \frac{1}{2}[L_\lambda : \alpha\alpha :] = \frac{1}{2} : [L_\lambda\alpha]\alpha : +\frac{1}{2} : \alpha[L_\lambda\alpha] : +\frac{1}{2}\int_0^\lambda [[L_\lambda\alpha]_\mu\alpha]\,d\mu.$$

Substitute (15.6) for $[L_\lambda\alpha]$:

$$[L_\lambda L] = \frac{1}{2}(: \partial\alpha\,\alpha : + : \alpha\,\partial\alpha :) + \lambda : \alpha\alpha : +\frac{1}{2}\int_0^\lambda [(\partial + \lambda)\alpha_\mu\alpha]\,d\mu$$

$$= \partial L + 2\lambda L + \frac{1}{2}\int_0^\lambda (\lambda - \mu)\mu\,d\mu.$$

Since $\frac{1}{2}\int_0^\lambda (\lambda - \mu)\mu\,d\mu = \frac{\lambda^3}{12}$, we conclude:

$$[L_\lambda L] = (\partial + 2\lambda)\,L + \frac{\lambda^3}{12}. \tag{15.7}$$

Note that the λ-bracket $[L_\lambda L]$ in this example has the same form as the λ-bracket of the Virasoro formal distribution (13.29) with central charge $C = 1$. We have thus given a simpler and more natural proof of Proposition 2.3 from Lecture 2.

Definition 15.1. A formal distribution L with values in a Lie superalgebra is called a *Virasoro formal distribution*, if

$$[L_\lambda L] = (\partial + 2\lambda)\,L + \frac{\lambda^3}{12}C,$$

where C is a central element.

A generalization of the previous example of a Virasoro formal distribution is given by

$$L^{(\gamma)} = \frac{1}{2} : \alpha\alpha : +\gamma\partial\alpha, \qquad \gamma \in \mathbb{C}. \tag{15.8}$$

By the sesquilinearity of the λ-bracket, $[\partial\alpha_\lambda\alpha] = -\lambda^2$, hence

$$[L^{(\gamma)}{}_\lambda\alpha] = (\lambda + \partial)\alpha - \gamma\lambda^2. \tag{15.9}$$

Again, by the sesquilinearity, $[L^{(\gamma)}{}_\lambda\partial\alpha] = (\lambda + \partial)^2\alpha - \gamma(\lambda + \partial)\lambda^2 = (\lambda + \partial)^2\alpha - \gamma\lambda^3$ (since $\partial\lambda = 0$). Also $\frac{1}{2}[L^{(\gamma)}{}_\lambda : \alpha\alpha :] = (\lambda + \frac{\partial}{2}) : \alpha\alpha : +\frac{\lambda^3}{12} - \gamma\lambda^2\alpha$. So

$$[L^{(\gamma)}{}_\lambda L^{(\gamma)}] = (\partial + 2\lambda)L^{(\gamma)} + \frac{\lambda^3}{12}c_\gamma, \qquad \text{where } c_\gamma = 1 - 12\gamma^2. \quad (15.10)$$

Hence, $L^{(\gamma)}$ is a Virasoro formal distribution with central charge c_γ.

15.3.   Virasoro formal distribution for neutral free fermions

Consider an odd one-dimensional vector superspace $A = \mathbb{C}\psi$ with the bilinear form $\langle \psi | \psi \rangle = 1$ and its Clifford affinization C_A (see Example 13.3). We put the central element $\mathbb{1}$ of C_A to be equal 1, which means that we work in the quotient of the universal enveloping algebra $U(C_A)$ by the ideal generated by $(\mathbb{1} - 1)$. The coefficients $\psi_{(n)} = \psi \otimes t^n$ of the formal distribution

$$\psi(z) \;\; = \;\; \sum_{n \in \mathbb{Z}} \psi_{(n)} z^{-n-1}$$

satisfy the commutation relations $[\psi_{(m)}, \psi_{(n)}] = \delta_{m+n,-1}$. To make the notation more symmetric, one writes this formal distribution in the form

$$\psi(z) \;\; = \;\; \sum_{n \in -\frac{1}{2} + \mathbb{Z}} \psi_n z^{-n-\frac{1}{2}},$$

where now the basis elements of C_A satisfy $[\psi_m, \psi_n] = \delta_{m,-n}$ (the deeper reasons of doing this will become clear when we introduce the notion of conformal weight in the next lecture). From (13.38) the λ-bracket is $[\psi_\lambda \psi] = 1$. Then by the sesquilinearity of the λ-bracket,

$$[\psi_\lambda \partial \psi] = (\partial + \lambda)1 = \lambda, \quad [\partial \psi_\lambda \psi] = -\lambda, \quad \text{and} \quad [\partial^2 \psi_\lambda \psi] = \lambda^2.$$

Consider the formal distribution

$$L(z) = \frac{1}{2} : \partial_z \psi(z)\psi(z) : . \quad (15.11)$$

By the non-commutative Wick formula,

$$\begin{aligned}
[\psi_\lambda L] \;\; &= \frac{1}{2}[\psi_\lambda : \partial \psi \, \psi :] \\
&= \frac{1}{2} : [\psi_\lambda \partial \psi] \, \psi : - \frac{1}{2} : \partial \psi \, [\psi_\lambda \psi] : + \frac{1}{2} \int_0^\lambda [[\psi_\lambda \partial \psi]_\mu \psi] \, d\mu \\
&= \frac{1}{2}\lambda\psi - \frac{1}{2}\partial\psi + 0 = \frac{1}{2}(-\partial + \lambda)\psi,
\end{aligned}$$

and by the skewcommutativity of the λ-bracket,

$$[L_\lambda \psi] = \left(\partial + \frac{1}{2}\lambda\right)\psi. \tag{15.12}$$

Note that by the quasicommutativity of the normal ordered product, $: \psi\psi :$
$=: \partial\psi\,\partial\psi := 0$, so, using the sesquilinearity of the λ-bracket and the non-commutative Wick formula, we get:

$$[L_\lambda L] = \frac{1}{2}[L_\lambda : \partial\psi\,\psi :]$$

$$= \frac{1}{2} : \left(\left(\partial + \frac{1}{2}\lambda\right)(\partial + \lambda)\psi\right)\psi : + \frac{1}{2} : (\partial\psi)\left(\partial + \frac{1}{2}\lambda\right)\psi :$$

$$+ \frac{1}{2}\int_0^\lambda [[L_\lambda\partial\psi]_\mu \psi]d\mu$$

$$= \frac{1}{2} : \left(\partial^2\psi + \frac{3\lambda}{2}\partial\psi + \frac{\lambda^2}{2}\psi\right)\psi : + \frac{1}{2}\left(: \partial\psi\,\partial\psi : + \frac{\lambda}{2} : \partial\psi\,\psi :\right)$$

$$+ \frac{1}{2}\int_0^\lambda \left([\partial^2\psi_\mu\psi] + \frac{3\lambda}{2}[\partial\psi_\mu\psi] + \frac{\lambda^2}{2}[\psi_\mu\psi]\right)d\mu$$

$$= (\partial + 2\lambda)L + \frac{1}{2}\int_0^\lambda \left(\mu^2 - \frac{3}{2}\lambda\mu + \frac{1}{2}\lambda^2\right)d\mu.$$

Since $\frac{1}{2}\int_0^\lambda \left(\mu^2 - \frac{3}{2}\lambda\mu + \frac{1}{2}\lambda^2\right)d\mu = \frac{\lambda^3}{24}$, we obtain:

$$[L_\lambda L] = (\partial + 2\lambda)L + \frac{\lambda^3}{24}. \tag{15.13}$$

Therefore, the formal distribution (15.11) is a Virasoro formal distribution
with central charge $C = \frac{1}{2}$. We have thus given a proof of Proposition 3.7
from Lecture 3 in the Neveu-Schwarz sector. The Ramond sector will be
discussed in the last section of Lecture 18, on twisted representations.

15.4. Virasoro formal distribution for charged free fermions

Let A be an odd two-dimensional vector superspace with a basis $\{\psi^+, \psi^-\}$
and with the skew-supersymmetric bilinear form with the matrix $\begin{pmatrix} 0 & 1 \\ 1 & 0 \end{pmatrix}$ in
this basis. Let C_A be the Clifford affinization of A, defined as in Example

13.3. As in §15.3, we put the central element $\mathbb{1}$ of C_A to be equal to 1, which means that we work in the quotient of the universal enveloping algebra $U(C_A)$ by the ideal generated by $(\mathbb{1} - 1)$. Let $\psi^{\pm}(z)$ be the formal distributions of free fermions that correspond to the elements $\psi^{\pm}$ of the basis of A. They are called *charged free fermions*. According to (13.38), the λ-brackets of $\psi^{\pm}(z)$ are

$$[\psi^{\pm}{}_{\lambda}\psi^{\mp}] = 1, \quad [\psi^{\pm}{}_{\lambda}\psi^{\pm}] = 0.$$

For any $\beta \in \mathbb{C}$ define a formal distribution

$$L^{(\beta)}(z) \quad = \quad \beta : \partial\psi^+ \psi^- : + (1 - \beta) : \partial\psi^- \psi^+ : . \tag{15.14}$$

Then $L^{(\beta)}(z)$ is a Virasoro formal distribution. Indeed, first let us prove that

$$[L^{(\beta)}{}_{\lambda}\psi^+] = (\partial + (1 - \beta)\lambda)\psi^+, \quad [L^{(\beta)}{}_{\lambda}\psi^-] = (\partial + \beta\lambda)\psi^-. \tag{15.15}$$

By the non-commutative Wick formula,

$$[\psi^{\pm}{}_{\lambda}L^{(\beta)}] = \beta[\psi^{\pm}{}_{\lambda} : \partial\psi^+\psi^- :] + (1 - \beta)[\psi^{\pm}{}_{\lambda} : \partial\psi^-\psi^+ :]$$

$$= \beta : [\psi^{\pm}{}_{\lambda}\partial\psi^+]\psi^- : - \beta : \partial\psi^+[\psi^{\pm}{}_{\lambda}\psi^-] :$$

$$+ \beta \int_0^{\lambda} [[\psi^{\pm}{}_{\lambda}\partial\psi^+]_{\mu}\psi^-]d\mu + (1 - \beta) : [\psi^{\pm}{}_{\lambda}\partial\psi^-]\psi^+ :$$

$$- (1 - \beta) : \partial\psi^-[\psi^{\pm}{}_{\lambda}\psi^+] :$$

$$+ (1 - \beta) \int_0^{\lambda} [[\psi^{\pm}{}_{\lambda}\partial\psi^-]_{\mu}\psi^+]d\mu.$$

Using that $[\psi^{\pm}{}_{\lambda}\partial\psi^{\mp}] = \lambda$, we get

$$[\psi^+{}_{\lambda}L^{(\beta)}] = (-\beta\partial + (1 - \beta)\lambda)\psi^+, \quad [\psi^-{}_{\lambda}L^{(\beta)}] = (\beta\lambda - (1 - \beta)\partial)\psi^-$$

and by the skewcommutativity,

$$[L^{(\beta)}{}_{\lambda}\psi^+] = (\partial + (1 - \beta)\lambda)\psi^+, [L^{(\beta)}{}_{\lambda}\psi^-] = (\partial + \beta\lambda)\psi^-. \tag{15.16}$$

Let us compute $[L^{(\beta)}{}_\lambda L^{(\beta)}]$. By the non-commutative Wick formula,

$$[L^{(\beta)}{}_\lambda : \partial\psi^+\psi^- :] = : [L^{(\beta)}{}_\lambda \partial\psi^+]\psi^- : + : \partial\psi^+[L^{(\beta)}{}_\lambda \psi^-] :$$
$$+ \int_0^\lambda [[L^{(\beta)}{}_\lambda \partial\psi^+]_\mu \psi^-]d\mu$$
$$= : (\lambda + \partial)(\partial + (1-\beta)\lambda)\psi^+\psi^- : + : \partial\psi^+(\partial + \beta\lambda)\psi^- :$$
$$+ \int_0^\lambda [(\lambda + \partial)(\partial + (1-\beta)\lambda)\psi^+{}_\mu \psi^-]d\mu.$$

Using the sesquilinearity of the λ-bracket, one can compute that

$$\int_0^\lambda [(\lambda + \partial)(\partial + (1-\beta)\lambda)\psi^+{}_\mu \psi^-]d\mu = -\frac{\lambda^3}{6} + \frac{\lambda^3}{2}(1-\beta). \qquad (15.17)$$

Similarly,

$$[L^{(\beta)}{}_\lambda : \partial\psi^-\psi^+ :] = : [L^{(\beta)}{}_\lambda \partial\psi^-]\psi^+ : + : \partial\psi^-[L^{(\beta)}{}_\lambda \psi^+] :$$
$$+ \int_0^\lambda [[L^{(\beta)}{}_\lambda \partial\psi^-]_\mu \psi^+]d\mu$$
$$= : (\lambda + \partial)(\partial + \beta\lambda)\psi^-\psi^+ : + : \partial\psi^-(\partial + (1-\beta)\lambda)\psi^+ :$$
$$+ \int_0^\lambda [(\lambda + \partial)(\partial + \beta\lambda)\psi^-{}_\mu \psi^+]d\mu.$$

and

$$\int_0^\lambda [(\lambda + \partial)(\partial + \beta\lambda)\psi^-{}_\mu \psi^+]d\mu = -\frac{\lambda^3}{6} + \frac{\lambda^3}{2}\beta. \qquad (15.18)$$

Then

$$[L^{(\beta)}{}_\lambda L^{(\beta)}] = \beta : ((\partial + \lambda)(\partial + (1-\beta)\lambda)\psi^+)\,\psi^- :$$
$$+ \beta : \partial\psi^+(\partial + \beta\lambda)\psi^- :$$
$$+ (1-\beta) : ((\partial + \lambda)(\partial + \beta\lambda)\psi^-)\,\psi^+ :$$
$$+ (1-\beta) : \partial\psi^-(\partial + (1-\beta)\lambda)\psi^+ : +\frac{\lambda^3}{12}c_\beta, \qquad (15.19)$$

where $c_\beta = -12\beta^2 + 12\beta - 2$. Using that $: \psi^+\psi^- : + : \psi^-\psi^+ : = 0$, which follows from the quasicommutativity of the normal ordered product, and that ∂_w is a derivation of the normal ordered product, formula (15.19) can be simplified to

$$[L^{(\beta)}_{\ \lambda}L^{(\beta)}] = (\partial + 2\lambda)L^{(\beta)} + \frac{\lambda^3}{12}c_\beta, \qquad c_\beta = -12\beta^2 + 12\beta - 2. \quad (15.20)$$

In other words, $L^{(\beta)}$ is a Virasoro formal distribution with central charge c_β.

Note that this c_β coincides with the one in §4.5. This is not a mere coincidence, as will be explained in Lecture 16 (see §16.4).

LECTURE 16

16.1. Conformal weights

In this lecture we introduce an important book-keeping device – the conformal weight, and the related notion of an energy operator. We return to the boson-fermion correspondence, discussed in Lectures 5 and 6, this time considered from the viewpoint of quantum fields, and as an application, we derive the classical Jacobi triple product identity.

Definition 16.1. We say that a formal $\mathfrak{g}$-valued distribution a is an eigendistribution of *conformal weight* $\Delta \in \mathbb{C}$ with respect to a Virasoro formal distribution L if

$$[L_\lambda a] = (\partial + \Delta\lambda)\, a + o(\lambda), \tag{16.1}$$

where $o(\lambda)$ denotes a polynomial divisible by λ^2. If $o(\lambda) = 0$, then a is said to be *primary*.

Example 16.1. By (15.6), the free boson $\alpha(z)$ is primary of conformal weight 1 with respect to $L(z)$. By (15.9), the same formal distribution $\alpha(z)$ has conformal weight 1 with respect to $L^{(\gamma)}$ from (15.8), but it is not primary.

From (15.15) the charged free fermions ψ^+ and ψ^- are primary with respect to $L^{(\beta)}$ from (15.14) of conformal weights $1 - \beta$ and β respectively.

By (15.12), the neutral free fermion ψ is primary of conformal weight $\frac{1}{2}$ with respect to L, defined in (15.11).

A Virasoro formal distribution has conformal weight 2 with respect to itself (and it is not primary unless the central charge is zero).

Proposition 16.1. Let a and b be formal distributions of conformal weights Δ_a and Δ_b. Then
(a) the formal distribution ∂a has conformal weight $\Delta_{\partial a} = \Delta_a + 1$;
(b) the formal distribution $a_{(n)}b$ has conformal weight $\Delta_a + \Delta_b - n - 1$;
(c) the normal ordered product $:ab:$ has conformal weight $\Delta_{:ab:} = \Delta_a + \Delta_b$.

Proof.   (a) From Proposition 13.5,

$$[L_\lambda \partial a] = (\partial + \lambda)[L_\lambda a] = (\partial + \lambda)\left((\partial + \Delta_a\lambda)a + o(\lambda)\right),$$
$$= (\partial + \Delta_a\lambda + \lambda)\,\partial a + o(\lambda).$$

(b) From (15.1),

$$[L_\lambda (a_{(n)}b)] = a_{(n)}[L_\lambda b] + \sum_{k\geq 0}\frac{\lambda^k}{k!}\,[L_\lambda a]_{(n+k)}b$$
$$= a_{(n)}(\partial b + \Delta_b\lambda b) + (\partial a + \Delta_a\lambda a)_{(n)}\,b + (\lambda\partial a)_{(n+1)}\,b\ + o(\lambda).$$

Using the sesquilinearity property (15.3), we obtain

$$[L_\lambda (a_{(n)}b)] = \partial(a_{(n)}b) + na_{(n-1)}b + \Delta_b\lambda a_{(n)}b - na_{(n-1)}b$$
$$+ \Delta_a\lambda a_{(n)}b - (n+1)\lambda\,a_{(n)}b + o(\lambda)$$
$$= (\partial + (\Delta_a + \Delta_b - n - 1)\lambda)\,a_{(n)}b + o(\lambda).$$

Assertion (c) follows from (b) by putting $n = -1$.   ∎

As we can see from Proposition 16.1, the conformal weight is a very convenient book-keeping device for the study of the OPE. Namely, if we prescribe to an eigendistribution the degree to be equal to its conformal weight, the degree of ∂ and of λ to be equal 1, then the degree of all terms in the λ-bracket of the eigendistributions a and b of conformal weights Δ_a and Δ_b respectively is equal to $\Delta_a + \Delta_b - 1$. For example, the degrees of all terms in $[L_\lambda L]$ equal 3. This observation puts a strong restriction on the possible form of an OPE.

Let $a(z)$ be a formal distribution of conformal weight Δ with respect to a Virasoro formal distribution $L(z)$. As we have seen in the case of the Virasoro formal distribution and shall see in many other examples, it is often convenient to write $a(z)$ in the form

$$a(z) = \sum_{n\in-\Delta+\mathbb{Z}} a_n z^{-n-\Delta}. \tag{16.2}$$

Then $a_{(n)} = a_{n-\Delta+1}$ and the commutator formula (13.17) for a local pair $(a(z), b(z))$ becomes

$$[a_m , b_n] = \sum_{j \geq 0} \binom{\Delta + m - 1}{j} (a_{(j)}b)_{m+n}. \tag{16.3}$$

Also

$$(\partial a)_n = -(n + \Delta)a_n. \tag{16.4}$$

In particular, if we put in (16.3) $a(z)$ to be a Virasoro formal distribution $L(z)$, and $b(z)$ to be an eigendistribution of conformal weight Δ_b with respect to $L(z)$, then, by the definition of the λ-bracket, we get

$$L_{(0)}b = \partial b, \quad L_{(1)}b = \Delta_b b, \tag{16.5}$$

and, since $\Delta_L = 2$, we have, by (16.3) and (16.4):

$$[L_m, b_n] = \binom{m + 1}{0}(L_{(0)}b)_{m+n} + \binom{m + 1}{1}(L_{(1)}b)_{m+n} + \dots$$

$$= (\partial b)_{m+n} + (m + 1)\Delta_b\, b_{m+n} + \dots.$$

Using (16.4), we get

$$[L_m, b_n] = (m(\Delta_b - 1) - n)\, b_{m+n} + \dots. \tag{16.6}$$

If $b(z)$ is primary with respect to $L(z)$, the rest of the terms in this equation is zero:

$$[L_m, b_n] = (m(\Delta_b - 1) - n)\, b_{m+n}.$$

For $m = 0$ there are no other terms in (16.6), hence we obtain

$$[L_0, b_n] = -n\, b_n. \tag{16.7}$$

The operator L_0 is called an energy operator.

16.2. Sugawara construction

In Lecture 10 we constructed representations of Vir from representations of the Kac-Moody affinization $\hat{\mathfrak{g}}'$ of a simple Lie algebra $\mathfrak{g}$. If $\{u_i\}$ ($i = 1, \dots, \dim \mathfrak{g}$) is a basis of a simple Lie algebra $\mathfrak{g}$, and $\{u^i\}$ ($i = 1, \dots, \dim \mathfrak{g}$)

is the corresponding dual basis with respect to an invariant nondegenerate symmetric bilinear form $(\cdot|\cdot)$ on $\mathfrak{g}$, the Casimir operator is defined by the formula $\Omega_0 = \sum_{i=1}^{\dim \mathfrak{g}} u_i u^i$. By formula (10.4) and Remark 10.2, the adjoint action of the Casimir operator on $\mathfrak{g}$ is multiplication by a scalar $2\,g$, where g is called the dual Coxeter number.

Fix $m \in \mathbb{C}$. Below the calculations are done in the quotient algebra $U = U(\hat{\mathfrak{g}}')/(M - m)$ of the universal enveloping algebra $U(\hat{\mathfrak{g}}')$ by the ideal generated by $(M - m)$, or rather its completion. Consider the currents $u_i(z)$ and $u^i(z)$ and put

$$T(z) = \frac{1}{2} \sum_{i=1}^{\dim \mathfrak{g}} : u_i(z)\, u^i(z) :, \quad L^{\mathfrak{g}}(z) = \frac{1}{m+g} T(z), \qquad (16.8)$$

where we assume that $m \neq -g$. Note that, writing $T(z) = \sum_n T_n z^{-n-2}$, the coefficients of the formal distribution $T(z)$ are given by formula (10.7).

Let $a(z)$ be any current. Then by the non-commutative Wick formula (15.2) and by (13.31),

$$[a_\lambda T] = \sum_i \left(\frac{1}{2} : [a_\lambda u_i] u^i : + \frac{1}{2} : u_i [a_\lambda u^i] : + \frac{1}{2} \int_0^\lambda [[a_\lambda u_i]_\mu u^i] d\mu \right)$$

$$= \sum_i \left(\frac{1}{2} \left(: [a, u_i] u^i : + : u_i [a, u^i] : \right) + \frac{\lambda m}{2} \left((a|u_i) u^i + u_i (a|u^i) \right) \right.$$

$$\left. + \frac{1}{2} \int_0^\lambda [[a, u_i], u^i] d\mu + \frac{1}{2} \int_0^\lambda m\mu ([a, u_i]|u^i) d\mu \right).$$

Recall that the Casimir operator commutes with elements of $\mathfrak{g}$ (Lemma 10.1):

$$[a, \Omega_0] = \sum_i ([a, u_i] u^i + u_i [a, u^i]) = 0, \ a \in \mathfrak{g}.$$

But $\sum_i \left(: [a, u_i] u^i : + : u_i [a, u^i] : \right)$ is a quantum field, all of whose coefficients have the form $\sum_i \left([a, u_i]_{(m)} u^i{}_{(n)} + (u_i)_{(m)} [a, u^i]_{(n)} \right)$. Hence, by the same argument as in the proof of Lemma 10.1 and Lemma 10.2, all these coefficients are 0, and $\sum_i (: [a, u_i] u^i : + : u_i [a, u^i] :) = 0$. Next, by the invariance of the form $(\cdot|\cdot)$ and by (10.3), $\sum_i ([a, u_i]|u^i) = \sum_i (a|[u_i, u^i]) = 0$. Furthermore, by (10.2), $\sum_i (a|u_i) u^i = \sum_i (a|u^i) u_i = a$, and by Lemma 10.3, $\sum_i [[a, u_i] u^i] = 2ga$. Thus, we obtain

$$[a_\lambda T] = (m + g)\lambda a, \quad [a_\lambda L^{\mathfrak{g}}] = \lambda a, \quad [L^{\mathfrak{g}}_\lambda a] = (\partial + \lambda)a. \qquad (16.9)$$

By the non-commutative Wick formula (15.2),

$$[L^{\mathfrak{g}}{}_\lambda L^{\mathfrak{g}}]$$

$$= \frac{1}{2(m+g)} \sum_i \left(:(\partial+\lambda)u_i\,u^i: + :u_i(\partial+\lambda)u^i: + \int_0^\lambda [[L^{\mathfrak{g}}{}_\lambda u_i]_\mu u^i]d\mu \right).$$

Since ∂ is a derivation of the normal ordered product, we have $:(\partial u_i)u^i:$ $+ :u_i(\partial u^i): = \partial(:u_iu^i:)$, and the integral can be computed using (13.31) and the sesquilinearity property of the λ-bracket:

$$\sum_i \int_0^\lambda [[L^{\mathfrak{g}}{}_\lambda u_i]_\mu u^i]d\mu = \sum_i \int_0^\lambda (-\mu+\lambda)([u_i,u^i] + \mu\,m(u_i|u^i))d\mu$$

$$= m\dim\mathfrak{g}\int_0^\lambda(-\mu+\lambda)\mu\,d\mu = \frac{1}{6}\lambda^3 m\dim\mathfrak{g}.$$

Hence,

$$[L^{\mathfrak{g}}{}_\lambda L^{\mathfrak{g}}] = (\partial+2\lambda)L^{\mathfrak{g}} + \frac{m\dim\mathfrak{g}}{12(m+g)}\lambda^3, \qquad (16.10)$$

and $L^{\mathfrak{g}}(z)$ is a Virasoro formal distribution with central charge $c(m) = \frac{m\dim\mathfrak{g}}{m+g}$. By (16.9), all currents $a(z)$ are primary of conformal weight 1.

Note that by formulas (10.13)–(10.14) in Lecture 10, the related representation of the Virasoro algebra with central charge $c(m)$ was defined through a restricted representation of $U(\hat{\mathfrak{g}}')$ with the central element M acting as a scalar operator $m\,I$, where $m \neq -g$.

16.3. Bosonization of charged free fermions

The *bosonization* of charged free fermions is the normal ordered product

$$\alpha = :\psi^+\psi^-:, \qquad (16.11)$$

where $\psi^\pm$ are as in §15.4. By the non-commutative Wick formula,

$$[\psi^\pm{}_\lambda \alpha] = \mp\psi^\pm, \quad [\alpha\,{}_\lambda\psi^\pm] = \pm\psi^\pm, \quad [\alpha_\lambda\alpha] = \lambda.$$

Note that the λ-bracket $[\alpha_\lambda\alpha]$ has the same form as the λ-bracket (14.12) of the free boson defined by the oscillator algebra $\mathcal{A}$ (Example 14.5). Let $L^{(\beta)}$ be a Virasoro formal distribution defined by (15.14). By the non-commutative Wick formula, the sesquilinearity of the λ-bracket and by (15.15), we have

$$[L^{(\beta)}_\lambda\alpha] = \, : [L^{(\beta)}_\lambda\psi^+]\psi^- : + : \psi^+[L^{(\beta)}_\lambda\psi^-] : + \int_0^\lambda [[L^{(\beta)}_\lambda\psi^+]_\mu\psi^-]d\mu$$

$$= \, : (\partial + (1-\beta)\lambda)\psi^+\psi^- : + : \psi^+(\partial + \beta\lambda)\psi^- : + \frac{\lambda^2}{2}(1-2\beta)$$

$$= \, : \partial\psi^+\psi^- : + : \psi^+\partial\psi^- : + \lambda : \psi^+\psi^- : + \frac{\lambda^2}{2}(1-2\beta).$$

Since ∂ is a derivation of the normal ordered product, we conclude

$$[L^{(\beta)}_\lambda\alpha] = (\partial + \lambda)\alpha + \frac{\lambda^2}{2}(1-2\beta). \tag{16.12}$$

Hence the formal distribution α has conformal weight 1 with respect to $L^{(\beta)}$ for any β. It is primary only for $\beta = 1/2$. By (15.15), the formal distributions $\psi^\pm$ are also primary, with respect to $L^{(1/2)}$, of conformal weight $1/2$.

Put (cf. §15.2) for α from (16.11)

$$L(z) = \frac{1}{2} : \alpha\alpha : . \tag{16.13}$$

By the results of §15.2, the quantum field L is a Virasoro formal distribution with central charge 1. Using the non-commutative Wick formula, we find

$$[\psi^\pm_\lambda L] = \frac{1}{2}[\psi^\pm_\lambda : \alpha\alpha :] = \frac{1}{2}\left(: [\psi^\pm_\lambda\alpha]\alpha : + : \alpha[\psi^\pm_\lambda\alpha] : + \int_0^\lambda [[\psi^\pm_\lambda\alpha]_\mu\alpha]d\mu \right)$$

$$= \mp\frac{1}{2}\left(: \psi^\pm\alpha : + : \alpha\psi^\pm : \right) + \frac{1}{2}\lambda\psi^\pm.$$

By the quasicommutativity of the normal ordered product (15.4), $: \alpha\psi^\pm : = : \psi^\pm\alpha : \pm\partial\psi^\pm$, hence

$$[\psi^\pm_\lambda L] = \mp : \psi^\pm\alpha : + \frac{1}{2}(-\partial + \lambda)\psi^\pm.$$

But, again, by the quasicommutativity of the normal ordered product, $2 : \psi^\pm \psi^\pm : = 0$. Hence, by quasiassociativity (17.29) (which will be proved later), we have:

$$: \psi^+ \alpha : = : \psi^+ : \psi^+ \psi^- ::$$

$$= :: \psi^+ \psi^+ : \psi^- : - \sum_{j=0}^\infty \left(\psi^+_{(-j-2)} (\psi^+_{(j)} \psi^-) - \psi^+_{(-j-2)} (\psi^+_{(j)} \psi^-) \right) = 0,$$

and similarly $: \psi^- \alpha : = : \psi^- : \psi^+ \psi^- :: = - : \psi^- : \psi^- \psi^+ :: = 0$. Hence, $[\psi^\pm_\lambda L] = \frac{1}{2}(-\partial + \lambda)\psi^\pm$, and, by the skewcommutativity of the λ-bracket, we obtain:

$$[L_\lambda \psi^\pm] = \left(\partial + \frac{1}{2}\lambda \right) \psi^\pm. \tag{16.14}$$

Alternatively, using the theory of vertex algebras, we will prove in §17.3 a simpler formula for L:

$$L(z) = L^{(1/2)} = \frac{1}{2} \left(: \partial \psi^+ \, \psi^- : + : \partial \psi^- \, \psi^+ : \right). \tag{16.15}$$

It follows from formula (15.16) that $\psi^\pm$ are primary quantum fields of conformal weight $1/2$ with respect to $L(z)$.

Expand the formal distributions α, $\psi^\pm$ and $L^{(1/2)}$ as in (16.2):

$$\psi^\pm(z) = \sum_{n \in -1/2 + \mathbb{Z}} \psi^\pm_n z^{-n-1/2},$$

$$\alpha(z) = \sum_{m \in \mathbb{Z}} \alpha_m z^{-m-1},$$

$$L^{(1/2)}(z) = \sum_{m \in \mathbb{Z}} L^{(1/2)}_m z^{-m-2}.$$

Then, by (16.3) and (16.4), the OPE can be written as commutation relations

$$[\psi^\pm_m, \psi^\pm_n] = 0, \qquad [\psi^\pm_m, \psi^\mp_n] = \delta_{m+n,0},$$
$$[\alpha_m, \alpha_n] = m\delta_{m+n,0}, \quad [\alpha_m, \psi^\pm_n] = \pm\psi^\pm_{m+n}.$$

In particular,

$$[\alpha_0, \psi^\pm_n] = \pm\psi^\pm_n. \tag{16.16}$$

Similarly, the λ-brackets of $L^{(1/2)}$ with α and $\psi^{\pm}$ correspond to commutation relations

$$[L_m^{(1/2)}, \alpha_n] = -n\alpha_{m+n} \quad (m, n \in \mathbb{Z}),$$
$$[L_m^{(1/2)}, \psi_n^{\pm}] = -\left(n + \frac{m}{2}\right)\psi_{m+n}^{\pm} \quad (m \in \mathbb{Z}, \ n \in -1/2 + \mathbb{Z}). \tag{16.17}$$

The commutation relations for the energy operator $L_0^{(1/2)}$ are as in (16.7):

$$[L_0^{(1/2)}, \alpha_n] = -n\alpha_n \quad (n \in \mathbb{Z}), \tag{16.18}$$

$$[L_0^{(1/2)}, \psi_n^{\pm}] = -n\psi_n^{\pm} \quad (n \in -1/2 + \mathbb{Z}). \tag{16.19}$$

16.4.   Irreducibility theorem for the charge decomposition

Let $F = F(A) = \operatorname{Ind}_{A[t] \oplus \mathbb{C}1}^{C_A} \mathbb{C}$ be the fermionic Fock representation defined in Example 14.4 for the two-dimensional odd superspace $A = \mathbb{C}\psi^+ + \mathbb{C}\psi^-$ of charged free fermions, as in §15.4 and §16.3. We denote by $|0\rangle$ the image of 1 in F and call it the vacuum vector. It follows from (14.13) that

$$\alpha_n |0\rangle = 0, \quad \text{if } n \geq 0, \qquad L_n^{(1/2)} |0\rangle = 0, \quad \text{if } n \geq 0. \tag{16.20}$$

By the PBW theorem (see §14.1), the vectors

$$\psi_{-j_t}^- \cdots \psi_{-j_1}^- \psi_{-i_s}^+ \cdots \psi_{-i_1}^+ |0\rangle \quad (0 \leq j_1 < \cdots < j_t, \ 0 \leq i_1 < \cdots < i_s) \tag{16.21}$$

form a basis of F. Using (16.19), (16.20), and (16.16) we obtain that the elements of the basis (16.21) are common eigenvectors of the operators $L_0^{(1/2)}$ and α_0 with the eigenvalues $\sum i_k + \sum j_l$, called *energy* and $m = s - t$, called *charge*, respectively.

Write the charge decomposition

$$F = \bigoplus_{m \in \mathbb{Z}} F^{(m)},$$

where $F^{(m)}$ is the α_0-eigenspace of eigenvalue (charge) m. Since the charge operator α_0 commutes with all α_n, it follows that $F^{(m)}$ is invariant under the action of all α_n, i.e. under the action of the oscillator algebra $\mathscr{A}$.

Define the *m-th charged vacuum* as the following element of $F^{(m)}$:

$$|m\rangle = \begin{cases} \psi^+_{-\frac{2m-1}{2}} \cdots \psi^+_{-\frac{1}{2}}|0\rangle, & \text{if } m > 0, \\ \psi^-_{-\frac{2|m|-1}{2}} \cdots \psi^-_{-\frac{1}{2}}|0\rangle, & \text{if } m < 0. \end{cases}$$

This vector has charge m and energy $m^2/2$. It is clear that this is a unique, up to a constant factor, vector in $F^{(m)}$ of lowest energy (equal $m^2/2$). Note also that $|m\rangle$ is a *singular* vector for the oscillator algebra $\mathscr{A}$, namely, one has: $\alpha_n|m\rangle = 0$ for $n > 0$, $\alpha_0|m\rangle = m|m\rangle$.

As we have seen in Lecture 5, the key fact behind the boson-fermion correspondence is the irreducibility of the representation of $\mathscr{A}$ in $F^{(m)}$. We proved this fact there by a combinatorial argument. Here we give a representation-theoretical proof.

Theorem 16.1. The representation of the oscillator algebra $\mathscr{A}$ in $F^{(m)}$ is irreducible.

Proof. Suppose that $F^{(m)}$ contains a proper nonzero subrepresentation W of $\mathscr{A}$. Then the energy operator $L_0 = L_0^{(1/2)}$ is a diagonal operator both on W and on $F^{(m)}/W$, and either W or $F^{(m)}/W$ does not contain the vector $|m\rangle$. We denote the one which does not contain $|m\rangle$ by M. Let $v \in M$ be a vector of minimal energy a, so that:

$$L_0 v = a v, \quad a > \frac{m^2}{2}.$$

On the other hand, $L_0 = \frac{\alpha_0^2}{2} + \sum_{k>0} \alpha_{-k}\alpha_k$ and by (16.18), vector $\alpha_k v$ has energy $a - k$, so $\alpha_k v = 0$ for $k > 0$. We conclude that

$$L_0 v = \frac{\alpha_0^2}{2}v = \frac{m^2}{2}v,$$

which leads to a contradiction. ∎

By Proposition 2.1 we get the following

Corollary 16.1. The representation of $\mathscr{A}$ in $F^{(m)}$ is equivalent to the bosonic Fock representation of charge m:

$$B^{(m)} = \mathbb{C}[x_1, x_2, \dots]$$

with

$$\alpha_0 = mI, \ \alpha_k = \frac{\partial}{\partial x_k}, \ \alpha_{-k} = kx_k \ (k > 0). \qquad \blacksquare$$

Recall that in §4.2 we introduced the representation of $\mathscr{A}$ in the infinite wedge space $\bigwedge^\infty V$ over a vector space V with a basis $\{v_i\}_{i\in\mathbb{Z}}$. It is immediate to see that $\bigwedge^\infty V$ can be canonically identified with F, by identifying the infinite monomial $v_0 \wedge v_{-1} \wedge \ldots$ with $|0\rangle$, the wedging operator $\hat{v}_n$, defined by (5.15), with the action of $\psi^+_{-n+1/2}$ on F, and the contraction operator $\check{v}_n^*$, defined by (5.16), with the action of $\psi^-_{n-1/2}$ on F. Then the shift operator $\hat{r}_m(\Lambda_k)$ from (5.18) is identified with the action of α_k ($k \neq 0$), and the semi-infinite monomial $v_m \wedge v_{m-1} \wedge \ldots$ is identified with $|m\rangle$. Finally, the generating series $X(u)$ and $X^*(u)$, defined by formula (5.21), get identified with the quantum fields $u\psi^+(u)$ and $\psi^-(u)$ respectively.

Hence, equation (7.2) can be written in the form

$$\sum_{k\in\mathbb{Z}} \psi^+_{-k+1/2}\tau \otimes \psi^-_{k-1/2}\tau = 0,$$

or, equivalently, in the form (cf. §7.3)

$$\text{Res}_z \left(\psi^+(z)\tau \otimes \psi^-(z)\tau\right) = 0.$$

This is the equation of the KP hierarchy in the fermionic picture.

In Lecture 4 the representation $\hat{r}$ of the Lie algebra $\mathfrak{a}_\infty$ was constructed. Formulas (4.51) that define the operators $\hat{r}(E_{ij})$ on F can be written in terms of the generating series $\sum_{i,j\in\mathbb{Z}} E_{ij}z^{i-1}w^{-j}$ as follows. By (4.51), (5.17), (5.19), we have, using the above identification:

$$\sum \hat{r}(E_{ij})z^{i-1}w^{-j} = \sum_{i>0,j\in\mathbb{Z}} \hat{r}(E_{ij})z^{i-1}w^{-j}$$

$$+ \sum_{i\leq 0, j\neq i} \hat{r}(E_{ij})z^{i-1}w^{-j} + \sum_{i\leq 0} \hat{r}(E_{ii})z^{i-1}w^{-i}$$

$$= \sum_{i>0,j\in\mathbb{Z}} \hat{v}_i\check{v}_j^* z^{i-1}w^{-j} + \sum_{i\leq 0, j\neq i} \hat{v}_i\check{v}_j^* z^{i-1}w^{-j} + \sum_{i\leq 0}(\hat{v}_i\check{v}_i^* - 1)z^{i-1}w^{-i}$$

$$= \sum_{i>0, j\in\mathbb{Z}} \hat{v}_i \check{v}_j^* z^{i-1} w^{-j} - \sum_{i\leq 0, j\in\mathbb{Z}} \check{v}_j^* \hat{v}_i z^{i-1} w^{-j}$$

$$= \sum_{i<0, j\in\mathbb{Z}} \psi^+_{i+1/2} \psi^-_{j+1/2} z^{-i-1} w^{-j-1} - \sum_{i\geq 0, j\in\mathbb{Z}} \psi^-_{j+1/2} \psi^+_{i+1/2} z^{-i-1} w^{-j-1}$$

$$= \; : \psi^+(z)\psi^-(w): \, ,$$

where, generalizing Definition 14.5, we let

$$: a(z)b(w) : \, = a(z)_+ b(w) + p(a,b)b(w)a(z)_- .$$

Next, we compare the coefficients of the Virasoro formal distribution $L^{(\beta)}$ defined by (15.14), acting on F, with the operators L_n, defined by (4.61). First, provided that $n \neq 1$, using (14.13) and (13.18), we have:

$$: \partial\psi^+\psi^- :_{(n)} = \sum_{j\in\mathbb{Z}} (-j)\psi^+_{j-1/2}\psi^-_{n-j-1/2} ,$$

and similarly,

$$: \partial\psi^-\psi^+ :_{(n)} = \sum_{j\in\mathbb{Z}} (-j)\psi^+_{n-j-1/2}\psi^-_{j-1/2} .$$

Hence for $n \neq 0$ we have:

$$L_n^{(\beta)} = \beta : \partial\psi^+\psi^- :_{(n+1)} + (1-\beta) : \partial\psi^-\psi^+ :_{(n+1)}$$

$$= \sum_{k\in\mathbb{Z}} (k + (1-\beta)(n+1))\psi^+_{-k-1/2}\psi^-_{n+k+1/2} .$$

By substitution of the index of summation k by $k - n - 1$ and identifying the action of $\psi^+_{-(k-n)+1/2}$ on F with $\hat{v}_{k-n}$ and the action of $\psi^-_{k-1/2}$ with $\check{v}_k^*$, the operator $L_n^{(\beta)}$ becomes

$$L_n^{(\beta)} = \sum_{k\in\mathbb{Z}} (k - \beta(n+1))\hat{v}_{k-n}\check{v}_k^*$$

$$= \sum_{k\in\mathbb{Z}} (k - \beta(n+1))\hat{r}(E_{k-n,k}) \quad (n \neq 0).$$

For $n = 0$ we have:

$$
\begin{aligned}
L_0^{(\beta)} &= \beta : \partial\psi^+\psi^- :_{(1)} + (1 - \beta) : \partial\psi^-\psi^+ :_{(1)} \\
&= \sum_{j \geq 0}(j + 1 - \beta)\psi^+_{-j-1/2}\psi^-_{j+1/2} + \sum_{j < 0}(-j - 1 + \beta)\psi^-_{j+1/2}\psi^+_{-j-1/2} \\
&= \sum_{j \geq 0}(j + 1 - \beta)\hat{v}_{j+1}\check{v}^*_{j+1} + \sum_{j < 0}(-j - 1 + \beta)\check{v}^*_{j+1}\hat{v}_{j+1} \\
&= \sum_{j \geq 0}(j + 1 - \beta)\hat{v}_{j+1}\check{v}^*_{j+1} + \sum_{j < 0}(j + 1 - \beta)(\hat{v}_{j+1}\check{v}^*_{j+1} - 1) \\
&= \sum_{j \geq 0}(j + 1 - \beta)\hat{r}(E_{j+1,j+1}) + \sum_{j < 0}(j + 1 - \beta)\hat{r}(E_{j+1,j+1}) \\
&= \sum_{j \in \mathbb{Z}}(j - \beta)\hat{r}(E_{jj}).
\end{aligned}
$$

Comparing the above formulas with formulas (4.11), (4.51) (5.17) and (4.61), we see that $L_n^{(\beta)}$ coincides with L_n, defined by (4.61), for all $n \in \mathbb{Z}$, provided that we let the parameter $\alpha = 0$ in (4.11).

16.5.   An application: the Jacobi triple product identity

Define the character of the representation of $\mathscr{A}$ on F as

$$
\mathrm{ch}\, F \;\; = \;\; \mathrm{tr}_F\, q^{L_0^{(1/2)}} z^{\alpha_0}.
$$

The form of the basis (16.21) shows that

$$
\mathrm{ch}\, F \;\; = \;\; \prod_{j=1}^{\infty}(1 + zq^{j-1/2})(1 + z^{-1}q^{j-1/2}). \tag{16.22}
$$

On the other hand, by the Irreducibility Theorem 16.1 and Proposition 2.1, the vectors

$$
\alpha_{-j_s}\ldots\alpha_{-j_1}|m\rangle \qquad (0 < j_1 \leq j_2 \leq \ldots) \tag{16.23}
$$

form a basis of $F^{(m)}$. The vector of the form (16.23) has energy $m^2/2 + j$, where $j = j_1 + \ldots + j_s$. Denote by $F_j^{(m)}$ the span of all vectors of charge m and energy $m^2/2 + j$. By (16.21), $\dim F_j^{(m)} = p(j)$, where $p(j)$ is the number of partitions of j in a sum of positive integers (the classical partition function). Hence

$$\operatorname{ch} F = \sum_{m \in \mathbb{Z}, j \in \mathbb{Z}_+} (\dim F_j^{(m)}) z^m q^{\frac{m^2}{2}+j}$$

$$= \sum_{m \in \mathbb{Z}, j \in \mathbb{Z}_+} p(j) z^m q^{\frac{m^2}{2}+j} = \sum_{m \in \mathbb{Z}} z^m \frac{q^{\frac{m^2}{2}}}{\prod_{j=1}^\infty (1-q^j)}. \quad (16.24)$$

Comparing (16.22) and (16.24) we get

$$\prod_{j=1}^\infty (1-q^j)(1+zq^{j-1/2})(1+z^{-1}q^{j-1/2}) = \sum_{m \in \mathbb{Z}} z^m q^{\frac{m^2}{2}}. \quad (16.25)$$

Replacing z by $-zq^{-\frac{1}{2}}$ in (16.25), we get the standard form of the *Jacobi triple product identity*:

$$\prod_{j=1}^\infty (1-q^j)(1-zq^{j-1})(1-z^{-1}q^j) = \sum_{m \in \mathbb{Z}} (-z)^m q^{m(m-1)/2}. \quad (16.26)$$

In a slightly different form this identity appeared in §11.2 as a special case of the Weyl-Macdonald-Kac formula. Replacing q in (16.26) by q^3 and then z by q^2, reduces (16.26) to the Euler identity (11.19), used in the proof of Proposition 11.3.

16.6. Restricted representations of free fermions

Let A be a finite-dimensional purely odd superspace endowed with a non-degenerate symmetric bilinear form $\langle \cdot | \cdot \rangle$, and let C_A be its Clifford affinization, defined in §13.4. Let $\{\psi^i\}_{i \in I}$ be an orthonormal basis of A. Then $\{\psi^i_n\}_{i \in I, n \in \frac{1}{2}+\mathbb{Z}} \cup \mathbb{1}$ is a basis of the Lie superalgebra C_A, the elements ψ^i_n being odd, the central element $\mathbb{1}$ even, and we have the following commutation relations:

$$[\psi^i_m, \psi^j_n] = \delta_{m,-n} \delta_{ij}\, \mathbb{1}. \quad (16.27)$$

It follows from §15.3 that the formal distributions $\psi^i(z) = \sum_{n \in \frac{1}{2}+\mathbb{Z}} \psi^i_n z^{-n-\frac{1}{2}}$ are primary eigendistributions of conformal weight $\frac{1}{2}$ with respect to the Virasoro formal distribution

$$L(z) = \frac{1}{2} \sum_i : \partial_z \psi^i(z) \psi^i(z) : = \sum_{n \in \mathbb{Z}} L_n z^{-n-2}.$$

Then we have

$$[L_0, \psi_n^i] = -n\psi_n^i. \tag{16.28}$$

Note also that

$$L_0 = \frac{1}{2} \sum_i \sum_{n>0} n\psi_{-n}^i \psi_n^i. \tag{16.29}$$

We call a representation of the Lie superalgebra C_A unital if $\mathbb{1}$ is represented by the identity operator. The proof of the following theorem uses the same idea as that of Theorem 16.1.

Theorem 16.2. The Lie superalgebra C_A has a unique, up to equivalence, unital restricted irreducible representation, and it is equivalent to $F(A)$ (see Example 14.4). Any unital restricted representation of C_A is equivalent to a direct sum of irreducible unital restricted representations.

Proof. Let V be a unital restricted representation of C_A. First we show that L_0 is diagonalizable on V with the spectrum in $\frac{1}{2}\mathbb{Z}_+$.

Let U^N denote the subalgebra of the universal enveloping superalgebra $U(C_A)$, generated by the elements $\psi_{-n}^i \psi_n^i$ with $i \in I$ and $0 < n \leq N$. Since V is a restricted representation, given $v \in V$, $\psi_n^i v = 0$ for all $i \in I$ and $n > N$ for N sufficiently large. Hence for such N, the subspace $U := U^N v$ is a finite-dimensional L_0-invariant subspace. We need to show that L_0 is diagonalizable on U. It suffices to show this for each summand $\psi_{-n}^i \psi_n^i$ of L_0. Let $v_1, \ldots, v_m$ be a basis of the kernel of ψ_n^i, acting on U. Since $(\psi_n^i)^2 = 0$, we see that $m \geq \frac{1}{2}\dim U$. But it is straightforward to show that the vectors $v_1, \ldots, v_m, \psi_{-n}^i v_1, \ldots, \psi_{-n}^i v_m$ are linearly independent. Hence they form a basis of U. Moreover, in this basis the matrix of ψ_n^i (respectively ψ_{-n}^i) is block-diagonal with blocks $\begin{pmatrix} 0 & 1 \\ 0 & 0 \end{pmatrix}$ (respectively, $\begin{pmatrix} 0 & 0 \\ 1 & 0 \end{pmatrix}$). It follows that in this basis the operator $\psi_{-n}^i \psi_n^i$ is diagonalizable with entries 0 or 1 on the diagonal. This implies the claim about L_0.

Let V_0 be the 0-th eigenspace of L_0 in V. We need to show that

$$V = U(C_A)V_0. \tag{16.30}$$

In the contrary case L_0 would be diagonalizable with strictly positive eigenvalues on $V/U(C_A)V_0$. Take in this space an eigenvector v with minimal eigenvalue of L_0. Then, by (16.28), $\psi_n^i v = 0$ for all $i \in I$, $n > 0$, hence, by (16.29), $L_0 v = 0$, which is a contradiction. Each nonzero vector $v \in V_0$ generates a subrepresentation, which is a quotient of $F(A)$. In fact, this is $F(A)$, since the representation of C_A in $F(A)$ is irreducible. Indeed, in the contrary case, as before, there would exist an eigenvector $v \in F(A)$ with positive eigenvalue, annihilated by all ψ_n^i with $n > 0$, hence $L_0 v = 0$ by (16.29), which gives a contradiction.

Hence, by (16.30), V is a direct sum of representations of C_A, equivalent to $F(A)$. ∎

LECTURE 17

17.1. Definition of a vertex algebra

The most natural framework for constructions of this book is the theory of vertex algebras and their representations. In a nutshell, a vertex algebra is a space of pairwise local quantum fields (see Definition 14.4), satisfying vacuum and translation covariance properties. These are the key axioms in Wightman's approach to quantum field theory (Wightman [1956]). Note that the Wightman axioms look different in that the state-field correspondence is replaced by the cyclicity of the vacuum vector. However, the Existence Theorem 17.1 shows that the latter property implies the former.

Definition 17.1. A *vertex algebra* $(V, |0\rangle, Y(a, z))$ is the following data:
– a vector superspace V (*the space of states*);
– an even vector $|0\rangle \in V$ (*the vacuum vector*);
– a linear parity-preserving map $a \mapsto Y(a, z)$ from the space V to the space of End V-valued quantum fields (called the *state-field correspondence*), such that the following axioms hold $(a, b \in V)$:

Vacuum Axioms

$$Y(|0\rangle, z) = I_V; \quad Y(a, z)|0\rangle = a + T(a)z + o(z),$$

where I_V is the identity operator on V, T is an even endomorphism of V, and $o(z)$ is a formal power series with coefficients in V, divisible by z^2.

Translation Covariance Axiom

$$[T, Y(a, z)] = \partial_z Y(a, z).$$

191

Locality Axiom

$$(z - w)^N [Y(a, z), Y(b, w)] = 0 \quad \text{for } N \in \mathbb{Z}_+ \text{ sufficiently large.}$$

The endomorphism $T \in \text{End}\, V$ is called the *infinitesimal translation operator*. The quantum fields $Y(a, z)$ are often called *vertex operators*, hence the name *vertex algebra*.

The original definition of a vertex algebra by Borcherds [1986] is different, but it is equivalent to this one (see §17.8).

Remark 17.1. Write $Y(a, z) = \sum_{n \in \mathbb{Z}} a_{(n)} z^{-n-1}$. Then for any $b \in V$ we have

$$Y(a, z)\, b = \sum_{n \in \mathbb{Z}} a_{(n)}(b) z^{-n-1}.$$

Thus the state-field correspondence is equivalent to giving for each $n \in \mathbb{Z}$ an n-th bilinear product on V by letting

$$a_{(n)} b := a_{(n)}(b).$$

We use the same notation for this operation as for the n-th product defined in (14.14), since we will prove later (Theorem 17.3) that the n-th products of quantum fields correspond to n-th products of states. In this form the vacuum axioms become $(a \in V)$:

$$|0\rangle_{(n)}\, a = \delta_{n,-1} a\, (n \in \mathbb{Z}), \quad a_{(n)} |0\rangle = \delta_{n,-1} a \quad (n \geq -1), \quad a_{(-2)} |0\rangle = Ta, \tag{17.1}$$

and the translation covariance axiom becomes $(a \in V)$:

$$[T, a_{(n)}] = -n a_{(n-1)} \qquad (n \in \mathbb{Z}). \tag{17.2}$$

Note that (17.1) implies the following very important property:

$$T|0\rangle = 0. \tag{17.3}$$

Finally, note that $a_{(n)} b = 0$ for $n \gg 0$, since $Y(a, z)$ is a quantum field. ∎

It is easy to construct a commutative vertex algebra, namely a vertex algebra satisfying the property $[Y(a, z), Y(b, w)] = 0$ for any $a, b \in V$:

Example 17.1. Let V be a unital commutative associative superalgebra with an even derivation T. Put

$$|0\rangle = 1,$$

$$Y(a,z)\, b = (e^{zT}a)b = \sum_{k \geq 0} \frac{z^k}{k!}(T^k a)b.$$

Then V is a vertex algebra. Indeed, the vacuum axioms hold:

$$Y(|0\rangle, z)\, b = (e^{zT}\, 1)b = b, \quad Y(a,z)\, |0\rangle = (e^{zT}\, a)1 = a + T(a)z + o(z),$$

the translation covariance axiom reduces to

$$[T, Y(a,z)]\, b = T((e^{zT}a)b) - (e^{zT}a)(Tb) = (e^{zT}Ta)\, b$$
$$= \partial_z(e^{zT}a)\, b = \partial_z Y(a,z)\, b,$$

and the locality follows from commutativity of the superalgebra V. Note that for $n \geq 0$ one has

$$a_{(n)}b = 0 \quad \text{and} \quad a_{(-1-n)}b = \frac{T^n(a)}{n!}b.$$

It is easy to show that any commutative vertex algebra is obtained in this way, so that commutative vertex algebras are in bijective correspondence with unital commutative associative differential algebras.

"Abstract" examples of vertex algebras are given by the following proposition:

Proposition 17.1. Let U^c be a completion of an associative algebra U, as in §14.1, and let V be a vector superspace of pairwise local continuous U^c-valued formal distributions. Assume that V is invariant under the derivation ∂_z, closed under all n-th products and contains 1, the identity element of U. Put $|0\rangle = 1$, and define the state-field correspondence on V by the formula

$$Y(a,z)\, b = \sum_{n \in \mathbb{Z}} (a_{(n)}b)z^{-n-1}, \qquad (17.4)$$

where $a_{(n)}b$ is the n-th product of formal distributions a and b in V given by (14.14). Then V is a vertex algebra, and $Ta(z) = \partial_z\, a(z)$.

Proof. Let us check the axioms of vertex algebra for $(V, |0\rangle, Y)$. The vacuum axioms in the form (17.1) follow from (14.14) $(a = a(w) \in V)$:

$$|0\rangle_{(n)} a = 1_{(n)}a = \delta_{n,-1}a \quad (n \in \mathbb{Z}),$$

and

$$a_{(n)}|0\rangle = a_{(n)}1 = \text{Res}_z\, a(z)(i_{z,w} - i_{w,z})(z - w)^n = \delta_{n,-1}a \quad (n \geq -1).$$

Since $Y(a, z)|0\rangle = a + T(a)z + o(z)$, we have

$$T(a) = a_{(-2)}|0\rangle = a_{(-2)}1 = :\partial a\, 1: = \partial a.$$

Note that for any formal distribution $b \in V$, by the sesquilinearity properties (15.3), we have

$$[T, a_{(n)}]b = \partial(a_{(n)}b) - a_{(n)}\partial b = -n a_{(n-1)}\, b,$$

and the translation covariance axiom in the from (17.2) follows.

To prove locality, note that from (17.4), using formula (14.14), for any formal distributions $a, b \in V$, one has

$$Y(a, x)\, b = \text{Res}_z\, \big(a(z)b(w)i_{z,w}\delta(z - w, x) - p(a, b)b(w)a(z)i_{w,z}\delta(z - w, x)\big).$$

Then for any formal distributions $a, b, c \in V$, one has

$$[Y(a, x), Y(b, y)]\, c(w)$$
$$= \text{Res}_{z_1} \text{Res}_{z_2} \big([a(z_1), b(z_2)]c(w)i_{z_1,w}i_{z_2,w}$$
$$- p(a, c)p(b, c)c(w)[a(z_1), b(z_2)]i_{w,z_1}i_{w,z_2}\big)\delta(z_1 - w, x)\delta(z_2 - w, y).$$

Since $(z_1 - z_2)^n[a(z_1), b(z_2)] = 0$ for $n \gg 0$, and since

$$(x - y) = (z_1 - z_2) - ((z_1 - w) - x) + ((z_2 - w) - y),$$

we get that for $n \gg 0$,

$$(x - y)^n[Y(a, x), Y(b, y)]c(w) = 0,$$

and the pair $(Y(a, x), Y(b, y))$ is local. ∎

Remark 17.2. Given a collection $\mathcal{F}$ of pairwise local End V-valued quantum fields, we may take its closure $\bar{\mathcal{F}}$, which is the minimal space of quantum fields, containing $\mathcal{F}$ and I_V, and which is closed under derivation ∂_z

and under all n-th products. By Dong's Lemma 15.1, the space $\bar{\mathcal{F}}$ of quantum fields is still local, hence, by Proposition 17.1, it is a vertex algebra. It follows that all identities, proved in this lecture for a vertex algebra, in particular, the n-th product identity, proved in §17.4, hold for quantum fields from $\mathcal{F}$. ∎

17.2. Existence Theorem

A construction of most of the important examples of (non-commutative) vertex algebras is given by the following

Theorem 17.1. (Existence Theorem) Let U be a unital associative super-algebra with a nested collection of subalgebras $\{U_i\}_{i \in I}$, satisfying properties (14.1) and (14.2). Let $\mathcal{F} = \{a^j(z)\}_{j \in J}$ be a collection of pairwise local continuous U-valued formal distributions. Assume that the superalgebra U has an even derivation $\tilde{T}$, such that

$$\tilde{T}a^j(z) = \partial_z a^j(z), \quad j \in J. \tag{17.5}$$

Denote by A the subalgebra of U, generated by all coefficients of the formal distributions $a^j(z)$. Let W be a $\tilde{T}$-invariant subalgebra of U, containing U_i for some $i \in I$, and assume that $U = AW$. Consider the representation V of U, induced from the trivial one-dimensional representation of W, i.e. $V = U/UW$ with the action of U defined by the left multiplication in U. Denote by $\bar{a}^j(z)$ the image of $a^j(z)$ in $(\text{End } V)[[z, z^{-1}]]$ under this representation.

Then V carries a structure of a vertex algebra, such that

(i) the vacuum vector $|0\rangle$ is the image of $1 \in U$ in V;

(ii) the infinitesimal translation operator $T \in \text{End } V$ is induced by $\tilde{T} \in \text{End } U$;

(iii) the state-field correspondence is given by the following well-defined map:

$$Y(\bar{a}^{j_1}_{(n_1)} \cdots \bar{a}^{j_k}_{(n_k)}|0\rangle, z) = \bar{a}^{j_1}(z)_{(n_1)}(\ldots (\bar{a}^{j_k}(z)_{(n_k)} I_V) \ldots).$$

Proof. We begin the proof with some simple observations. First, by Proposition 14.1, the representation of U on V is restricted, hence the formal distributions from the collection $\mathcal{F}$ are represented by quantum fields. The same holds for n-th products of these formal distributions by Proposition 14.2(b). Denote by $\mathcal{B}$ the span of formal distributions of the form

$$a^{j_1}(z)_{(n_1)}(\ldots(a^{j_{k-1}}(z)_{(n_{k-1})}(a^{j_k}(z)_{(n_k)}I_V))\ldots), \qquad (17.6)$$

where $j_1, \ldots, j_k \in J$, $n_1, \ldots, n_k \in \mathbb{Z}$. Then all formal distributions from $\mathcal{B}$ are represented by End V-valued quantum fields. All these quantum fields are pairwise local by Dong's lemma.

Next, since $\tilde{T}$ is a derivation of the algebra U, it induces a derivation of all n-th products on the space of all (continuous) U^c-valued formal distributions. It follows from (17.5) that

$$\tilde{T}b(z) = \partial_z b(z) \quad \text{for all} \quad b(z) \in \mathcal{B}.$$

Denote by $\bar{\mathcal{B}}$ the image of $\mathcal{B}$ in (End $V)[[z, z^{-1}]]$ under the representation of U in V, and by $\bar{a}$ the quantum field, which is the image of $a \in \mathcal{B}$. Since obviously $T|0\rangle = 0$, it follows that

$$[T, \bar{b}(z)] = \partial_z \bar{b}(z) \quad \text{for} \quad \bar{b}(z) \in \bar{\mathcal{B}}. \qquad (17.7)$$

Here we used that ∂_z is a derivation of all n-th products (Proposition 14.2(c)), and that $ad\,T$ is a derivation of all n-th products of quantum fields.

Next we need three lemmas. In all these lemmas V is a vector superspace, $|0\rangle \in V$ is an even vector, and T is an even endomorphism of V, such that $T|0\rangle = 0$.

Lemma 17.1. If $b(z)$ is a quantum field, such that

$$[T, b(z)] = \partial_z b(z), \qquad (17.8)$$

then $b(z)|0\rangle \in V[[z]]$. Moreover,

$$b(z)|0\rangle = e^{zT}b, \quad \text{where} \quad b = b(z)|0\rangle|_{z=0}. \qquad (17.9)$$

Proof. We have (cf. (17.2)): $[T, b_{(n)}] = -nb_{(n-1)}$. Apply both sides to $|0\rangle$ to obtain:

$$Tb_{(n)}|0\rangle = -nb_{(n-1)}|0\rangle. \qquad (17.10)$$

Since $b_N|0\rangle = 0$ for some $N \geq 0$, we conclude from (17.10) that $b_{n-1} = 0$ if $n > 0$, which proves the first claim. In order to prove (17.9), note that both sides are formal power series in z with coefficients in V, which are equal at $z = 0$ and satisfy the same differential equation $\partial_z x(z) = T(x(z))$. ∎

Lemma 17.2. If two $\text{End}\,V$-valued quantum fields $a(z) = \sum_n a_{(n)} z^{-n-1}$ and $b(z) = \sum_n b_{(n)} z^{-n-1}$ satisfy the property that $a_{(n)}|0\rangle = 0$, $b_{(n)}|0\rangle = 0$ for $n \geq 0$, then, for any $k \in \mathbb{Z}$, $a(w)_{(k)}b(w)\,|0\rangle$ is a formal power series in w with the constant term

$$a(w)_{(k)}b(w)\,|0\rangle\,|_{w=0} = a_{(k)}(b_{(-1)}|0\rangle). \tag{17.11}$$

Proof. Recall that

$$a(w)_{(k)}b(w)\,|0\rangle$$
$$= \text{Res}_z\, i_{z,w}(z-w)^k a(z)b(w)|0\rangle - p(a,b)\text{Res}_z\, i_{w,z}(z-w)^k b(w)a(z)|0\rangle.$$

Since $a_{(n)}|0\rangle = 0$ for $n \geq 0$, the second summand involves only non-negative powers of z, hence $\text{Res}_z\, i_{w,z}(z-w)^k b(w)a(z)|0\rangle = 0$. Therefore $a(w)_{(k)}b(w)\,|0\rangle$ contains no negative powers of w and

$$a(w)_{(k)}b(w)\,|0\rangle\,|_{w=0} = \text{Res}_z\, i_{z,w}(z-w)^k a(z)b(w)|0\rangle\,|_{w=0}$$
$$= \text{Res}_z\, z^k a(z)(b_{(-1)}|0\rangle) = a_{(k)}(b_{(-1)}|0\rangle). \qquad ∎$$

Lemma 17.3. Let $\mathcal{E}$ be a collection of pairwise local $\text{End}\,V$-valued quantum fields, such that (17.8) holds for any $b(z) \in \mathcal{E}$. By Lemma 17.1, we can define the map $f : \mathcal{E} \to V$ by

$$f(b(z)) = b(z)|0\rangle|_{z=0} \quad \text{for all} \quad b(z) \in \mathcal{E}.$$

Suppose that the map f is surjective. Then the map f is injective.

Proof. Suppose that $b(z) \in \text{Ker}\, f$. We have, by locality, for each $a(z) \in \mathcal{E}$ and some $N \in \mathbb{Z}_+$:

$$(z-w)^N b(z)a(w)|0\rangle = \pm(z-w)^N a(w)b(z)|0\rangle.$$

By (17.9), this can be rewritten as

$$(z-w)^N b(z)e^{wT}a = \pm(z-w)^N a(w)e^{zT}b,$$

where $a = f(a(z))$ and $b = f(b(z)) = 0$. Hence $(z - w)^N b(z) e^{wT} a = 0$. Letting $w = 0$ and canceling out z^N, we obtain $b(z)a = 0$. Since f is surjective, we conclude that $b(z) = 0$.   ∎

Lemma 17.1 allows us to construct the map

$$\varphi : \mathcal{B} \to V, \quad b(z) \mapsto \bar{b}(z)|0\rangle|_{z=0}. \tag{17.12}$$

By Lemma 17.2, the image of the formal distribution (17.6) under the map φ is the element $\bar{a}^{j_1}_{(n_1)} \dots \bar{a}^{j_k}_{(n_k)} |0\rangle$. Since $U = AW$, we conclude that the map φ is surjective. Lemma 17.3 implies that the map φ factors through to a bijective map $\bar{\varphi} : \bar{\mathcal{B}} \to V$. Hence we can invert the map $\bar{\varphi}$ to get a well-defined bijective map from V to a space of pairwise local quantum fields as in (iii) of Theorem 17.1. By (17.7) we have translation covariance, and by (17.9) we have the second of the vacuum axioms, while the first one is obvious. This completes the proof of Theorem 17.1.   ∎

Remark 17.3. Suppose that all the elements $\{a^j_{(-n)}\}_{j \in J, n \geq 1}$ are linearly independent, and suppose, moreover, that there exists an ordering of these elements: $v_1, v_2, \dots$, such that the ordered monomials (14.3) form a basis of $U \bmod W$. Then the map (17.12) is bijective.   ∎

17.3.   Examples of vertex algebras

By the Existence Theorem, all restricted representations V^c, $V^m(\mathfrak{g})$, $B^{(0)}$, and $F(A)$, constructed in Examples 14.1-14.4 carry a canonical structure of a vertex algebra. Indeed, all the conditions of this theorem hold if we take for U the universal enveloping algebra of the corresponding Lie superalgebra, for $\mathcal{F}$ the families of formal distributions, described in Examples 13.1-13.3, for $\tilde{T}$ the derivation ad L_{-1} in the case of the Virasoro algebra, and $(-\partial_z)$ in all other cases, and for W the subalgebra of U, generated respectively by:

− $\{L_n\}_{n \geq -1}$ and $C - c$ for the Virasoro algebra;

− $\mathfrak{g}[t]$ and $M - m$ for the Kac-Moody affinization;

− $\{a_n\}_{n \geq 0}$ and $\mathbb{1} - 1$ for the free boson;

− $A[t]$ and $\mathbb{1} - 1$ for free fermions.

Note that $B^{(\mu)}$ with $\mu \neq 0$ does not carry a structure of a vertex algebra since the subalgebra of U generated by $a_0 - \mu$, a_1, a_2, $\ldots$ is not $\tilde{T}$-invariant.

Explicitly, by Theorem 17.1, the state-field correspondence in all these examples looks as follows.

Vectors $L_{-n_s-2} \ldots L_{-n_1-2}|0\rangle$ with $0 \leq n_1 \leq n_2 \leq \ldots$ form a basis of V^c, and we have:

$$Y(L_{-n_s-2} \ldots L_{-n_1-2}|0\rangle, z) = \frac{: \partial_z^{n_s} L(z) \ldots \partial_z^{n_1} L(z) :}{n_s! \ldots n_1!}. \tag{17.13}$$

Here and further on the normal ordered product is taken from right to left, e.g. $: abc :$ stands for $: a : bc :::$.

Next, vectors of the form $a^{i_s}_{(-n_s-1)} \ldots a^{i_1}_{(-n_1-1)}|0\rangle$ with $a^{i_1}, \ldots, a^{i_s} \in \mathfrak{g}$ and $0 \leq n_1 \leq n_2 \leq \ldots$ span $V^m(\mathfrak{g})$, and we have:

$$Y(a^{i_s}_{(-n_s-1)} \ldots a^{i_1}_{(-n_1-1)}|0\rangle, z) = \frac{: \partial_z^{n_s} a^{i_s}(z) \ldots \partial_z^{n_1} a^{i_1}(z) :}{n_s! \ldots n_1!}. \tag{17.14}$$

In the special case of the free boson, vectors $a_{(-n_s-1)} \ldots a_{(-n_1-1)}|0\rangle$ with $0 \leq n_1 \leq n_2 \leq \ldots$ form a basis of $B^{(0)}$ and we have:

$$Y(a_{-n_s-1} \ldots a_{-n_1-1}|0\rangle, z) = \frac{: \partial_z^{n_s} a(z) \ldots \partial_z^{n_1} a(z) :}{n_s! \ldots n_1!}. \tag{17.15}$$

Finally, in the case of free fermions, consider the most important case when A is purely odd. Choosing a basis $\phi^1, \ldots, \phi^d$ of A, we obtain the following basis of $F(A)$:

$$(\phi^d_{(-n_s-1)} \ldots \phi^d_{(-n_1-1)}) \ldots (\phi^1_{(-m_s-1)} \ldots \phi^1_{(-m_r-1)})|0\rangle$$

with $0 \leq n_1 \cdots < n_s$, $0 \leq m_1 < \cdots < m_r$, and we have:

$$Y(\phi^d_{(-n_s-1)} \ldots \phi^d_{(-n_1-1)} \ldots \phi^1_{(-m_s-1)} \ldots \phi^1_{(-m_r-1)}|0\rangle, z)$$
$$= \frac{: \partial_z^{n_s} \phi^d(z) \ldots \partial_z^{n_1} \phi^d(z) \ldots \partial_z^{m_r} \phi^1(z) \ldots \partial_z^{m_1} \phi^1(z) :}{(n_s! \ldots n_1!) \ldots (m_r! \ldots m_1!)}. \tag{17.16}$$

Some important special cases of (17.13)-(17.16) are

$$Y(L_{-2}|0\rangle, z) = L(z); \quad Y(a_{(-1)}|0\rangle, z) = a(z), a \in \mathfrak{g};$$
$$Y(\phi_{(-1)}|0\rangle, z) = \phi(z), \phi \in A. \tag{17.17}$$

Remark 17.4. The *lattice vertex algebra* V_Q, where Q is an integral lattice with bilinear form $(\cdot|\cdot)$ (see Kac [1998]) can be constructed along the same lines. The corresponding superalgebra U is defined as follows. Let ε : $Q \times Q \to \mathbb{C}^\times$ be a two-cocycle, satisfying the condition

$$\varepsilon(\alpha, \beta) = (-1)^{(\alpha|\alpha)(\beta|\beta)+(\alpha|\beta)} \varepsilon(\beta, \alpha),$$

and let $\mathbb{C}_\varepsilon[Q]$ be the group superalgebra of Q with the twisted product $e^\alpha e^\beta = \varepsilon(\alpha, \beta) e^{\alpha+\beta}$ and the parity $p(e^\alpha) = (\alpha|\alpha) \bmod 2$. Let $\mathfrak{h} = \mathbb{C} \otimes_{\mathbb{Z}} Q$, viewed as an abelian Lie algebra. Then U is the twisted tensor product of the algebra $U(\hat{\mathfrak{h}}')$ and the superalgebra $\mathbb{C}_\varepsilon[Q]$ with the relations

$$h_{(n)} e^\alpha = \delta_{n,0}(\alpha|h) e^\alpha + e^\alpha h_{(n)}, \quad h \in \mathfrak{h}, \ \alpha \in Q.$$

The collection of subalgebras of U with properties (14.1), (14.2) are those of $U(\hat{\mathfrak{h}}')$, tensored with $1 \in \mathbb{C}_\varepsilon[Q]$, and the subalgebra W is that of $U(\hat{\mathfrak{h}}')$, tensored with 1. The derivation $\tilde{T}$ is extended from $U(\hat{\mathfrak{h}}')$ by letting $\tilde{T}(e^\alpha) = \alpha e^\alpha$. Finally, the collection of U-valued (rather U^c-valued) formal distributions is obtained by adding to those for $U(\hat{\mathfrak{h}}')$ the vertex operators, similar to those that appear in the boson-fermion correspondence (which is the case of $Q = \mathbb{Z}$, see Lecture 5). ■

Example 17.2. The simplest and the most important examples of vertex algebras $F(A)$ of free fermions are the following.

(a) $A = \mathbb{C}\psi$ as in §15.3. Then the vertex algebra $F(A)$ is called the *neutral* (or Neveu-Schwarz) *free fermion*. In this case formula (15.11) gives rise to the Neveu-Schwarz representations of the Virasoro algebra, constructed in §3.4.

(b) $A = \mathbb{C}\psi^+ + \mathbb{C}\psi^-$ as in §15.4. The vertex algebra $F(A)$ is called *charged free fermions*. Let us give a simple proof of formula (16.15), using that the state-field correspondence s is bijective. Indeed, the state $s(L(z))$ corresponding to the field $L(z)$ is $\frac{1}{2}\alpha^2_{-1} = \frac{1}{2}\alpha_{-1}(\psi^+_{-\frac{1}{2}}\psi^-_{-\frac{1}{2}}|0\rangle)$. Since $\alpha_{-1} = \sum_{j \in \frac{1}{2}+\mathbb{Z}} \psi^+_{-j}\psi^-_{j-1}$, one easily derives: $s(L(z)) = \frac{1}{2}(\psi^+_{-\frac{3}{2}}\psi^-_{-\frac{1}{2}} + \psi^-_{-\frac{3}{2}}\psi^+_{-\frac{1}{2}})|0\rangle$, proving the claim.

The boson-fermion correspondence, discussed in Lecture 5, basically says that the vertex algebra $F(A)$ from (b) is isomorphic to the lattice vertex algebra $V_{\mathbb{Z}}$ (cf. Remark 17.4).

Remark 17.5. In Lecture 13 we have seen that all Lie superalgebras, which produced the above examples of vertex algebras, give rise to Lie conformal superalgebras. This is not a coincidence. Namely, to each Lie conformal superalgebra $\mathcal{R}$ one can canonically associate a Lie superalgebra

Lie $\mathcal{R}$, using formulas (13.17) and (13.18), a derivation $\tilde{T}$, given by $\tilde{T}(a_{(n)}) = -na_{(n-1)}$, and a collection of pairwise local formal distributions $\mathcal{F} = \{a(z) = \sum_n a_{(n)} z^{-n-1}\}_{a\in\mathcal{R}}$. Then, taking for U the universal enveloping superalgebra of *Lie* $\mathcal{R}$ and for W its subalgebra spanned by all $\{a_{(n)}\}_{a\in\mathcal{R}, n\geq 0}$, we construct, according to the Existence Theorem, the vertex algebra $V(\mathcal{R})$, called the *universal enveloping vertex algebra* of the Lie conformal superalgebra $\mathcal{R}$. One can find details and more examples in Kac [1998] and De Sole-Kac [2006]. ∎

In view of this remark, one calls the vertex algebra V^c the *universal Virasoro vertex algebra* of central charge c, and $V^m(\mathfrak{g})$ the *universal affine vertex algebra* of level m.

17.4. Uniqueness Theorem and n-th product identity

Theorem 17.2. (Uniqueness Theorem) Let V be a vertex algebra and let $B(z)$ be an $\operatorname{End} V$-valued quantum field, such that the pairs $(B(z), Y(a,z))$ are local for all $a \in V$. Assume that $B(z)|0\rangle = Y(b,z)|0\rangle$ for some $b \in V$. Then $B(z) = Y(b,z)$.

Proof. Let $B_1(z) = B(z) - Y(b,z)$. Then $B_1(z)|0\rangle = 0$ and all the pairs $(B_1(z), Y(a,z))$ are local. Hence for each $a \in V$ there exists $N \in \mathbb{Z}_+$, such that $(z-w)^N B_1(z) Y(a,w)|0\rangle = \pm(z-w)^N Y(a,w) B_1(z)|0\rangle = 0$. Therefore, by (17.9), $(z-w)^N B_1(z) e^{wT} a = 0$. Letting $w = 0$, we get $B_1(z)a = 0$. ∎

In order to apply the Uniqueness Theorem to the proof of the important n-th product identity, we need the following lemma.

Lemma 17.4. Let $(V, |0\rangle, Y)$ be a vertex algebra. Then

$$(Y(a,z)_{(n)} Y(b,z))\,|0\rangle = Y(a_{(n)} b, z)|0\rangle. \tag{17.18}$$

Proof. By the vacuum axioms and Lemma 17.2, both sides of (17.18) are formal power series in z with the same constant term $a_{(n)}b$. Then lemma follows if we prove that both sides of (17.18) satisfy the same differential equation $\partial_z x(z) = Tx(z)$. The translation covariance axiom and (17.3) imply that the right-hand side of (17.18) satisfies this differential equation. As for the left-hand side of (17.18), recall that both operators, ∂_z and T, are derivations of n-th products. Then again, from the translation covariance axiom we obtain:

$$\partial_z \left(Y(a,z)_{(n)} Y(b,z) \right) |0\rangle = (\partial_z \, Y(a,z) \,_{(n)} \, Y(b,z)) |0\rangle$$
$$+ \, (Y(a,z)_{(n)} \, \partial_z \, Y(b,z)) |0\rangle$$
$$= (TY(a,z))_{(n)} Y(b,z) |0\rangle + Y(a,z)_{(n)} TY(b,z) |0\rangle$$
$$= T(Y(a,z)_{(n)} Y(b,z)) |0\rangle. \qquad \blacksquare$$

Theorem 17.3. For any $a, b \in V$ and for any $n \in \mathbb{Z}$ we have the following *n-th product identity*:

$$Y(a,z)_{(n)} Y(b,z) = Y(a_{(n)} b, z). \qquad (17.19)$$

Proof. By Dong's lemma and by the Uniqueness Theorem, it suffices to check that both sides, applied to the vacuum vector, are equal. But this follows from Lemma 17.4. $\blacksquare$

Observe that by letting $n = -1$ in Theorem 17.3, we obtain the (-1)st product identity:

$$Y(a_{(-1)} b, z) = \, : Y(a,z) Y(b,z) : ,$$

and letting $n = -2$ and $b = |0\rangle$, we obtain

$$Y(Ta, z) = \partial_z \, Y(a, z). \qquad (17.20)$$

Finally, using Proposition 13.3, we derive from Theorem 17.3 the following OPE for vertex operators:

$$[Y(a,z), Y(b,w)] = \sum_{j \geq 0} Y(a_{(j)} b, w) \partial_w^j \, \delta(z,w)/j!, \qquad (17.21)$$

or, equivalently, the following commutator formula for any $a, b \in V$ (cf. (13.17)):

$$[a_{(m)} \, , \, b_{(n)}] = \sum_{j \geq 0} \binom{m}{j} \left(a_{(j)} b \right)_{(m+n-j)}. \qquad (17.22)$$

17.5. Some constructions

The most popular examples of vertex algebras are the ones obtained from the universal enveloping vertex algebras of Lie conformal superalgebras and from lattice vertex algebras by some simple constructions, which we now describe (see Kac [1998] for more details).

(a) A *subalgebra* of a vertex algebra V is a subspace U of V, which contains $|0\rangle$, and such that

$$U_{(n)}U \subset U \quad \text{for all} \quad n \in \mathbb{Z}.$$

From the axioms of a vertex algebra it follows that a subalgebra U is a vertex algebra with the state-field correspondence $Y(a, z) = \sum_n a_{(n)}|_U z^{-n-1}$, $a \in U$.

(b) An *ideal* of a vertex algebra V is a subspace J that does not contain $|0\rangle$ and such that

$$V_{(n)}J \subset J \quad \text{and} \quad J_{(n)}V \subset J \quad \text{for all} \quad n \in \mathbb{Z}.$$

(Note that $TJ \subset J$.) It follows from the skewcommutativity of the λ-bracket and the quasicommutativity of the normal ordered product, that, equivalently, an ideal J is a T-invariant subspace of V that does not contain $|0\rangle$ and which is a left ideal, i.e. $V_{(n)}J \subset J$ for all $n \in \mathbb{Z}$.

The factor-space V/J of V by an ideal J has the canonical structure of a vertex algebra and we have a canonical homomorphism $V \to V/J$ of vertex algebras.

(c) The *tensor product* $U \otimes V$ of vertex algebras U and V is again a vertex algebra, defined as follows. The space of states is the tensor product $U \otimes V$ of vector superspaces, the vacuum vector is the tensor product of vacuum vectors $|0\rangle \otimes |0\rangle$, and the state-field correspondence is defined by

$$Y(v \otimes v, z) = Y(u, z) \otimes Y(v, z) = \sum_{m,n \in \mathbb{Z}} (u_{(m)} \otimes v_{(n)})z^{-m-n-2}.$$

(d) A *derivation* of a vertex algebra V of parity $\gamma \in \mathbb{Z}/2\mathbb{Z}$ is an endomorphism D of the vector superspace V of parity γ, such that

$$D(a_{(n)}b) = (Da)_{(n)}b + (-1)^{p(a)\gamma}a_{(n)}(Db) \quad \text{for any} \quad a, b \in V, n \in \mathbb{Z}.$$

Obviously the kernel of any even or odd derivation is a subalgebra of the vertex algebra V.

Letting $m = 0$ in (17.22) and applying both sides to a vector c, we see that the endomorphism $a_{(0)}$ is a derivation of the vertex algebra V for any

$a \in V$; it is called an *inner derivation*, associated to a. Consequently, given a subspace A of V, the intersection of kernels of endomorphisms $\{a_{(0)} | a \in A\}$ is a subalgebra of the vertex algebra V for any $a \in V$. Some vertex algebras called W-algebras, which are important both in mathematics and physics, are constructed in this way.

(e) The *set of fixed points* of any automorphism of a vertex algebra V is a subalgebra of V. If G is a group of automorphisms of a vertex algebra V, the set V^G of invariants of the G-action on V is a subalgebra of V. If G is finite, this subalgebra is called an *orbifold* in the physics literature.

(f) The *centralizer* of a subspace U of a vertex algebra V, defined as

$$C(U) = \{b \in V \mid [b_\lambda U] = 0\},$$

is a subalgebra of V. This follows from Proposition 15.1 and the n-th product formula. Some special cases of this construction are called *coset models* in the physics literature (cf. §10.2).

17.6.   Energy-momentum fields

Definition 17.2. An even vector ν in a vertex algebra V is called a *conformal vector*, if the corresponding field $Y(\nu, z) = \sum_{n \in \mathbb{Z}} L_n z^{-n-2}$ is a Virasoro field with central charge c that satisfies the properties:
(i) $L_{-1} = T$;
(ii) the operator L_0 is diagonalizable in V.

The number c is called *central charge* of ν, the vertex algebra V is called a *conformal* vertex algebra of rank c, and the quantum field $Y(\nu, z)$ is called the *energy-momentum field* of the vertex algebra V. As in §16.1, the operator L_0 is called the *energy operator*, and for its eigenvector a with eigenvalue Δ it is convenient to write the corresponding quantum field in the form

$$Y(a, z) = \sum_{n \in -\Delta + \mathbb{Z}} a_n z^{-n-\Delta}. \tag{17.23}$$

The following proposition allows one to construct automorphisms of vertex algebras.

Proposition 17.2. Let V be a vertex algebra with an energy operator L_0 all of whose eigenspaces are finite-dimensional. Let $a \in V$ be an eigenvector of L_0 with eigenvalue 1. Then the series $e^{a_0} = \sum_{n \in \mathbb{Z}_+} \frac{1}{n!} a_0^n$ converges to an automorphism of the vertex algebra V.

Proof. Since $\Delta_a = 1$, $a_0 = a_{(0)}$, hence a_0 is a derivation of the vertex algebra V. Since by (16.7), $[L_0, a_0] = 0$, all eigenspaces of L_0 are a_0-invariant and, since they are finite dimensional, the series e^{a_0} converges on each of them. It is an automorphism of the vertex algebra V since a_0 is its derivation. ∎

The vertex algebras V^c, $F(A)$, $V^m(\mathfrak{g})$, and $B^{(0)}$, constructed in §17.3, are conformal vertex algebras with conformal vectors described below.

The vertex algebra V^c has a conformal vector

$$\nu = L_{-2}|0\rangle. \tag{17.24}$$

Indeed, by (17.13), $Y(\nu, z) = L(z)$, where $L(z)$ is the Virasoro formal distribution defined in Example 13.1. Also, by construction of the state-field correspondence, $T = \text{ad}\, L_{-1}$, and since $[L_0, L_n] = -n L_n$ and $L_0 |0\rangle = 0$, the operator L_0 is diagonal in the basis consisting of monomials $L_{-n_s-2} \ldots L_{-n_1-2}|0\rangle$.

The vertex algebra of charged free fermions $F(A)$, where A is two-dimensional with basis $\{\psi^{\pm}\}$ as in §15.4, has a family of conformal vectors depending on the parameter β:

$$\nu^{\beta} = \beta \psi^+_{(-2)} \psi^-_{(-1)}|0\rangle + (1 - \beta) \psi^-_{(-2)} \psi^+_{(-1)}|0\rangle.$$

By (17.16), the field $Y(\nu^{\beta}, z)$ is a Virasoro formal distribution $L^{(\beta)}$ defined by (15.14).

In order to prove that the vector ν^{β} is a conformal vector we need the following simple lemma. The same lemma works for the next example of the Sugawara construction and for the example of the free boson.

Lemma 17.5. Let ν be a vector of a vertex algebra V, such that $L(z) = Y(\nu, z)$ is a Virasoro formal distribution, and let $\mathcal{F} = \{a^j\}_{j \in J}$ be a collection of vectors in V, such that the quantum fields $Y(a^j, z) = \sum_{n \in -\Delta_j + \mathbb{Z}} a^j_n z^{-n-\Delta_j}$ are eigendistributions for $L(z)$ with eigenvalues Δ_j, i.e. (16.1) holds for $a = a^j$, $\Delta = \Delta_j$. Assume that V is generated by $\mathcal{F}$, i.e. V is spanned by the vectors

$$a^{j_1}_{n_1} \ldots a^{j_s}_{n_s}|0\rangle, \quad j_1, \ldots, j_s \in J, \quad n_1, \ldots, n_s \in \mathbb{Z}. \tag{17.25}$$

Then ν is a conformal vector and the vector (17.25) is an eigenvector of L_0 with eigenvalue $-(n_1 + \ldots + n_s)$.

Proof.   We have, by the definition of the λ-bracket:

$$L_{-1}a^j = Ta^j, \; L_0a^j = \Delta_j a^j.$$

Since both $L_{-1} = L_{(0)}$ and T are derivations of n-th products and both annihilate $|0\rangle$, we deduce that $L_{-1} = T$. Since, by (16.7), $[L_0, a_n^j] = -na_n^j$ and also $L_0|0\rangle = 0$, we conclude that L_0 is diagonalizable, vectors (17.25) being its eigenvectors.   ∎

Let $\{u_i\}$ and $\{u^i\}$ be a basis and its dual basis of $\mathfrak{g}$, as in §16.2, where $\mathfrak{g}$ is either simple or abelian Lie algebra. The conformal vector of the vertex algebra $V^m(\mathfrak{g})$ is

$$\nu = \frac{1}{2(m+g)} \sum_i u_{i(-1)} u^i_{(-1)} |0\rangle, \tag{17.26}$$

where g is a dual Coxeter number ($g = 0$ if $\mathfrak{g}$ is abelian) and we assume that $m \neq -g$. By (17.14), the field $Y(\nu, z)$ is the Virasoro formal distribution $L^{\mathfrak{g}}(z)$ defined by (16.8), for which the currents $u_i(z)$ are eigendistributions. Since the vectors $u_{i(-1)}|0\rangle$ generate the vertex algebra $V^m(\mathfrak{g})$, by Lemma 17.5 we conclude that ν is its conformal vector.

A special case of the previous example is the following conformal vector of the vertex algebra $B^{(0)}(\mathfrak{g})$:

$$\nu = \frac{1}{2}\alpha^2_{(-1)}|0\rangle.$$

17.7.   Poisson like definition of a vertex algebra

Define a λ-bracket on V by

$$[a_\lambda b] = \sum_{n \in \mathbb{Z}_+} \frac{\lambda^n}{n!}(a_{(n)}b). \tag{17.27}$$

The n-th product identities (17.19) for $n \geq 0$ say that the λ-bracket of states and the λ-bracket of the associated vertex operators correspond to each other, and formula (17.20) says that the action of T on a state corresponds to the action of ∂_z on the associated vertex operator. Hence Proposition

13.5 implies that the λ-bracket (17.27) on V satisfies all the Lie conformal superalgebra axioms with $\partial = T$.

Next, in view of the (-1)st product identity (i.e. (17.19) with $n = -1$), it is natural to denote the (-1)st product on V by $: ab :$, and call it the normal ordered product.

The (-1)st product identity and Proposition 15.2 imply the quasicommutativity of the normal ordered product:

$$: ab : - p(a, b) : ba : = \int_{-T}^{0} [a_\lambda b] \, d\lambda. \qquad (17.28)$$

Computing the constant term in the (-1)st product identity, applied to $c \in V$, we obtain the *quasiassociativity*

$$:: ab : c : - : a : bc :: = \sum_{j=0}^{\infty} (a_{(-j-2)}(b_{(j)}c) + p(a, b)b_{(-j-2)}(a_{(j)}c)). \quad (17.29)$$

We observe that the associator of the normal ordered product, i.e., the expression $:: ab : c : - : a : bc ::$ is unchanged, up to the sign $p(a, b)$, if we exchange a and b. Such a product is called *left-symmetric*.

Finally, again, by the n-th product identity (17.19), the normal ordered product and the λ-bracket on V are related by the non-commutative Wick formula (cf. (15.2)):

$$[a_\lambda : bc :] = : [a_\lambda b] c : + p(a, b) : b [a_\lambda c] : + \int_{0}^{\lambda} [[a_\lambda b]_\mu c] \, d\mu. \qquad (17.30)$$

We thus arrive at a "Poisson like" definition of a vertex algebra. Namely, a vertex algebra is a vector superspace V with a vacuum vector $|0\rangle$ and an endomorphism T, endowed with two structures:

(i) a λ-bracket $[a_\lambda b]$, satisfying all the axioms of a Lie conformal superalgebra with $\partial = T$,

(ii) a normal ordered product $: ab :$, which defines a unital (with the identity element $|0\rangle$), differential (with the derivation T) quasicommutative left-symmetric superalgebra.

The two structures are related by the non-commutative Wick formula.

We just proved that the axioms of a vertex algebra imply all the above properties. The converse is more difficult and its proof can be found in Bakalov-Kac [2003].

Note that in a "quasiclassical limit" the λ-bracket still satisfies the axioms of a Lie conformal superalgebra, but the normal ordered product "degenerates" to a unital commutative associative product with a derivation, and the non-commutative Wick formula degenerates to the usual Leibniz-Wick formula (the "quantum corrections" disappear). The resulting algebraic structure is called a Poisson vertex algebra. It is very important for the theory of Hamiltonian PDE (see Barakat-De Sole-Kac [2009]).

17.8.   Borcherds identity

Proposition 17.3. (Borcherds identity) Let V be a vertex algebra. For any $k \in \mathbb{Z}$ and any $a, b \in V$ one has:

$$Y(a,z)Y(b,w)i_{z,w}(z-w)^k - p(a,b)Y(b,w)Y(a,z)i_{w,z}(z-w)^k \quad (17.31)$$
$$= \sum_{j \in \mathbb{Z}_+} Y(a_{(k+j)}b, w)\partial_w^j \delta(z,w)/j!.$$

Proof.   Denote the left-hand side of (17.31) by $d(z,w)$. It is a local formal distribution in z and w, hence by the Decomposition Theorem 13.1 we have:

$$d(z,w) = \sum_{j \in \mathbb{Z}_+} c^j(w)\partial_w^j \delta(z,w)/j!,$$

where

$$\begin{aligned}
c^j(w) &= \text{Res}_z\,(z-w)^j d(z,w) \\
&= \text{Res}_z\,(Y(a,z)Y(b,w)i_{z,w}(z-w)^{k+j} \\
&\quad - p(a,b)Y(b,w)Y(a,z)i_{w,z}(z-w)^{k+j}) \\
&= Y(a,w)_{(k+j)}Y(b,w) = Y(a_{(k+j)}b,w). \qquad \blacksquare
\end{aligned}$$

Equating the coefficients of $z^{-m-1}w^{-n-1}$ of both sides of (17.31), we obtain the following equivalent form of the Borcherds identity:

$$\sum_{j \in \mathbb{Z}_+} (-1)^j \binom{k}{j} \left(a_{(m+k-j)}b_{(n+j)} - (-1)^k p(a,b)b_{(n+k-j)}a_{(m+j)} \right) \quad (17.32)$$

$$= \sum_{j \in \mathbb{Z}_+} \binom{m}{j} (a_{(k+j)}b)_{(m+n-j)}.$$

Note that the Borcherds identity along with

$$Y(|0\rangle, z) = I, \quad a_{(-1)}|0\rangle = a \qquad (17.33)$$

provide an equivalent definition of a vertex algebra V. Indeed, since $Y(a, w)$ is a quantum field, we have $a_{(k+j)}b = 0$ for all $j \in \mathbb{Z}_+$ if k is sufficiently large. Hence the right-hand side of (17.31) vanishes for these k, therefore the locality axiom holds. Furthermore, the substitution $b = |0\rangle$ and $k = 0$ in (17.31) gives that $a_{(j)}|0\rangle = 0$ for all $j \geq 0$, so the vacuum axioms (17.1) follow. Next, let $Ta = a_{(-2)}|0\rangle$ and apply both parts of (17.32) with $m = 0$, $n = -2$ to $|0\rangle$. Using that $b_{(j)}|0\rangle = 0$ for $j \geq 0$, (17.32) reduces to

$$a_{(k)}b_{(-2)}|0\rangle - ka_{(k-1)}b_{(-1)}|0\rangle = (a_{(k)}b)_{(-2)}|0\rangle,$$

and from the second equality of (17.33), we obtain the translation covariance axiom (17.2).

LECTURE 18

18.1. Definition of a representation of a vertex algebra

Representation theory of vertex algebras is a rapidly developing field, which has its origin in conformal field theory. Roughly speaking, the representations of a vertex algebra are the building blocks of the quantum fields of a 2-dimensional conformal field theory. Here we just sketch some of the important ideas of the subject.

Recall that a representation of a unital associative algebra A in a vector space M is a linear map $A \to \operatorname{End} M$, $a \mapsto a^M$, such that $1 \mapsto I_M$ and the associativity axiom holds: $(ab)^M = a^M b^M$. A representation of a vertex algebra is defined similarly.

Definition 18.1. A *representation* of a vertex algebra V in a vector superspace M is a parity preserving linear map from V to the superspace of $\operatorname{End} M$-valued quantum fields:

$$a \mapsto Y^M(a, z), \quad Y^M(a, z) = \sum_{n \in \mathbb{Z}} a^M_{(n)} z^{-n-1},$$

such that (cf. (17.33), (17.31))

$$Y^M(|0\rangle, z) = I_M, \tag{18.1}$$

and Borcherds identity holds for any $k \in \mathbb{Z}$ and $a, b \in V$:

$$Y^M(a, z) Y^M(b, w) i_{z,w}(z - w)^k - p(a, b) Y^M(b, w) Y^M(a, z) i_{w,z}(z - w)^k$$
$$= \sum_{j \in \mathbb{Z}_+} Y^M(a_{(k+j)} b, w) \partial^j_w \delta(z, w) / j!. \tag{18.2}$$

211

Example 18.1. The state-field correspondence defines a representation of a vertex algebra on itself. It is called the *standard* or *adjoint representation*.

Equating the coefficients of $z^{-m-1}w^{-n-1}$ on both sides of (18.2), we obtain the following equivalent form of it $(a, b \in V, m, n, k \in \mathbb{Z})$:

$$\sum_{j\in\mathbb{Z}_+} (-1)^j \binom{k}{j} \left(a^M_{(m+k-j)} b^M_{(n+j)} - (-1)^k p(a, b) b^M_{(n+k-j)} a^M_{(m+j)} \right)$$

$$= \sum_{j\in\mathbb{Z}_+} \binom{m}{j} (a_{(k+j)} b)^M_{(m+n-j)}. \qquad (18.3)$$

A special case of (18.2) or (18.3) for $k = 0$ is the commutator formula (cf. (17.22)):

$$[a^M_{(m)}, b^M_{(n)}] = \sum_{j\in\mathbb{Z}_+} \binom{m}{j} (a_{(j)} b)^M_{(m+n-j)}, \qquad (18.4)$$

and, in particular, the locality of all pairs $(Y^M(a, z), Y^M(b, z))$.

Taking Res_z of both sides of (18.2), we obtain

$$Y^M(a, w)_{(k)} Y^M(b, w) = Y^M(a_{(k)} b, w), \quad k \in \mathbb{Z}, \qquad (18.5)$$

which for $k = -1$ gives

$$Y^M(a_{(-1)} b, z) = : Y^M(a, z) Y^M(b, z) : . \qquad (18.6)$$

Letting $b = |0\rangle$ and $k = -2$ in (18.5) gives

$$Y^M(Ta, z) = \partial_z Y^M(a, z). \qquad (18.7)$$

It is easy to show by looking at the proof of Proposition 17.3 that, conversely, identities (18.4), (18.6) and (18.7) imply the Borcherds identity (18.2). Similarly, the locality of all pairs $(Y^M(a, z), Y^M(b, z))$ and (18.5) imply (18.2). Hence we have the following equivalent definition of a representation of a vertex algebra in a vector superspace M: the vacuum vector goes to the identity operator on M, and the n-th product of any two elements of V goes to the n-th product of the corresponding $\mathrm{End}\, M$-valued local quantum fields for each $n \in \mathbb{Z}$.

18.2. Representations of the universal vertex algebras

Theorem 18.1. Let V be one of the vertex algebras V^c, $V^m(\mathfrak{g})$, $B^{(0)}$, and $F(A)$. Then there is a canonical bijective correspondence between representations of V and restricted representations of Lie algebras $\mathfrak{a} = Vir$, $\hat{\mathfrak{g}}'$, $\mathscr{A}$ and C_A respectively (see Examples 13.1-13.3), such that $C = cI$ for Vir, $M = mI$ for $\hat{\mathfrak{g}}'$, and $\mathbb{1} = I$ for $\mathscr{A}$ and C_A.

Proof. Consider a representation of V in M, $a \mapsto Y^M(a, z) = \sum_{n \in \mathbb{Z}} a_{(n)}^M z^{-n-1}$. Since $Y^M(a, z)$ is a quantum field, for any $v \in M$ we have $a_{(n)}v = 0$ if n is sufficiently large. Hence, due to (18.4), we obtain a restricted representation of the Lie algebra $\mathfrak{a}$ in M. For example, in the case of the Virasoro algebra, (18.4) shows that the coefficients of the quantum field $L^M(z) = Y^M(L_{-2}|0\rangle, z)$ satisfy the commutation relations of the Virasoro algebra, while in the case of the Kac-Moody affinization $\hat{\mathfrak{g}}'$, the coefficients of quantum fields $a^M(z) = Y^M(a_{(-1)}|0\rangle, z)$, $a \in \mathfrak{g}$, satisfy the commutation relations of $\hat{\mathfrak{g}}'$ (cf. (17.17)). Thus, any representation of the vertex algebra V in a vector space M is automatically a restricted representation of the corresponding Lie algebra $\mathfrak{a}$ with the corresponding value of the central element.

Conversely, consider a restricted representation of the Lie algebra $\mathfrak{a}$ in a vector space M, such that $C = cI$, $M = mI$, $\mathbb{1} = I$, and $\mathbb{1} = I$, respectively. Why can it be extended to a representation of the corresponding vertex algebra?

First, this representation extends to the universal enveloping algebra $U = U(\mathfrak{a})$, and, being restricted, it extends to the completion U^c. Second, it is easy to check that the conditions of Remark 17.3 hold in all cases. Hence the map $\phi : \mathcal{B} \to V$, defined by (17.12), can be inverted to give the inverse map

$$\phi^{-1} : V \to \mathcal{B} \subset U^c[[z, z^{-1}]].$$

Composing this map with the representation of U^c in M, we obtain the desired representation of V in M. ∎

Thus, any restricted representation M of the universal vertex algebra V restricts to a restricted representation of the Lie algebra $\mathfrak{a}$ with the corresponding eigenvalue of the central element. Conversely, any such representation M of $\mathfrak{a}$ extends uniquely to a representation of V.

18.3.   On representations of simple vertex algebras

A vertex algebra V is called *simple* if it contains no non-trivial ideals, or, equivalently, no T-invariant left ideals different from zero and V (see §17.5). But a left ideal of V is a subspace U, which is $a_{(n)}$-invariant for all $a \in V$ and $n \in \mathbb{Z}$. Since the representation of the oscillator algebra $\mathscr{A}$ in $B^{(0)}$ and that of the Clifford affinization C_A in $F(A)$ are irreducible, it follows that both $B^{(0)}$ and $F(A)$ are simple vertex algebras.

Proposition 18.1. Any irreducible representation of the vertex algebra $B^{(0)}$ (respectively, $F(A)$) is equivalent to the extension of the representation of the Lie algebra $\mathscr{A}$ in $B^{(\mu)}$, $\mu \in \mathbb{C}$ (respectively, of the Lie superalgebra C_A in $F(A)$).

Proof. It is easy to deduce from Proposition 2.1 that any restricted irreducible representation of $\mathscr{A}$ is equivalent to $B^{(\mu)}$ for some $\mu \in \mathbb{C}$, and likewise, any restricted irreducible representation of C_A is equivalent to $F(A)$. But a restricted representation of the Lie algebra $\mathfrak{a}$ is irreducible if and only if it is irreducible with respect to the completion U^c of the universal enveloping algebra $U(\mathfrak{a})$. Hence, by Theorem 18.1, Proposition 18.1 follows. ■

In sharp contrast to the vertex algebras of free bosons and free fermions, the vertex algebras V^c and $V^m(\mathfrak{g})$ are not always simple since the corresponding representations of Vir and $\hat{\mathfrak{g}}'$ are not always irreducible, as has been explained in Examples 14.1 and 14.2.

It follows from the proof of Proposition 3.3(c) that the representation of Vir in V^c has a unique maximal proper subrepresentation, which we denote by J^c. Since $T = L_{-1}$, the subspace J^c is automatically T-invariant, hence J^c is a unique maximal ideal of the vertex algebra V^c. Consequently, $V_c := V^c/J^c$ is the unique simple quotient of the vertex algebra V^c, called the *simple Virasoro vertex algebra of central charge c*.

Similarly, the universal affine vertex algebra $V^m(\mathfrak{g})$ of level m, where $\mathfrak{g}$ is a simple Lie algebra, has a unique maximal ideal, provided that $m + g \neq 0$. Indeed, in this case $V^m(\mathfrak{g})$ has the conformal vector ν, defined by (17.26), so that $Y(\nu, z) = \sum_{n \in \mathbb{Z}} L_n z^{-n-2}$ is an energy-momentum field. In particular, L_0 is diagonalizable, hence $V^m(\mathfrak{g})$ has a unique proper maximal $\hat{\mathfrak{g}}'$-invariant subspace, which we denote by $J^m(\mathfrak{g})$. Since $L_{-1} = T$, this subspace is T-invariant and therefore is the unique maximal ideal of the vertex algebra

$V^m(\mathfrak{g})$. The quotient vertex algebra $V_m(\mathfrak{g}) := V^m(\mathfrak{g})/J^m(\mathfrak{g})$ is called the *simple affine vertex algebra of level m.*

The following proposition is obvious.

Proposition 18.2. Any representation M of a factor vertex algebra V/J is obtained from a representation M of V with the trivial action of J, i.e. $Y^M(a, z) = 0$ for all $a \in J$. ∎

18.4. On representations of simple affine vertex algebras

Let $\mathfrak{g}$ be a simple finite-dimensional Lie algebra, and let e_i, f_i, h_i ($i = 1, \ldots, r = \text{rank } \mathfrak{g}$) be its Chevalley generators. Recall that the elements h_i form a basis of a Cartan subalgebra $\mathfrak{h}$ of $\mathfrak{g}$, the elements e_i (resp. f_i) generate a maximal nilpotent subalgebra $\mathfrak{n}_+$ (resp. $\mathfrak{n}_-$) of $\mathfrak{g}$, and we have the triangular decomposition $\mathfrak{g} = \mathfrak{n}_- \oplus \mathfrak{h} \oplus \mathfrak{n}_+$. Let θ be the highest root of $\mathfrak{g}$, so that a root vector e_θ (resp. $e_{-\theta}$) spans the center of $\mathfrak{n}_+$ (resp. of $\mathfrak{n}_-$). We choose the symmetric invariant bilinear form $(\cdot|\cdot)$ on $\mathfrak{g}$, normalized by the condition $(\theta|\theta) = 2$, and choose root vectors e_θ and $e_{-\theta}$, such that $(e_\theta|e_{-\theta}) = 1$. Since the restriction of the form $(\cdot|\cdot)$ to $\mathfrak{h}$ is nondegenerate, we can identify $\mathfrak{h}$ and $\mathfrak{h}^*$ via this form.

Let $\hat{\mathfrak{g}}'$ be the Kac-Moody affinization of the pair $(\mathfrak{g}, (\cdot|\cdot))$, and let $e_0 = e_{-\theta}t$, $h_0 = M - \theta$, $f_0 = e_\theta t^{-1}$. Then the elements e_i, f_i, h_i ($i = 0, 1, \ldots, r$) are Chevalley generators of the Lie algebra $\hat{\mathfrak{g}}'$, namely, they generate the Lie algebra $\hat{\mathfrak{g}}'$ and satisfy the Chevalley relations ($i, j = 0, \ldots, r$):

$$[h_i, h_j] = 0, \quad [e_i, f_j] = \delta_{ij}h_i, \quad [h_i, e_j] = a_{ij}e_j, \quad [h_i, f_j] = -a_{ij}e_j, \quad (18.8)$$

where (a_{ij}) is the extended Cartan matrix of $\mathfrak{g}$. The elements h_i ($i = 0, \ldots, r$) form a basis of the abelian subalgebra $\hat{\mathfrak{h}}'$, the elements e_i (resp. f_i) ($i = 0, \ldots, r$) generate a subalgebra $\hat{\mathfrak{n}}_+$ (resp. $\hat{\mathfrak{n}}_-$), and we have the triangular decomposition

$$\hat{\mathfrak{g}}' = \hat{\mathfrak{n}}_- \oplus \hat{\mathfrak{h}}' \oplus \hat{\mathfrak{n}}_+.$$

Note that a construction of the triangular decomposition for $\hat{sl}'_n$ can be obtained from that for $\hat{sl}_n$ in §9.4, by replacing $\hat{\mathfrak{h}} = \hat{\mathfrak{h}}' + \mathbb{C}d$ by $\hat{\mathfrak{h}}'$.

Fix $m \in \mathbb{Z}_+$ and consider the universal affine vertex algebra $V^m(\mathfrak{g})$. Denote $v_m := f_0^{m+1}|0\rangle$. The key observation is that

$$v_m \in J^m(\mathfrak{g}). \tag{18.9}$$

As usual, this is established by showing that the vector v_m is *singular*, i.e.

$$h_i v_m \in \mathbb{C} v_m \quad \text{and} \quad e_i v_m = 0 \quad \text{for all} \quad i = 0, \ldots, r, \tag{18.10}$$

which is easily deduced from the Chevalley relations (18.8). For example, $e_i v_m = f_0^{m+1} e_i |0\rangle = 0$ for $i \geq 1$, since $[e_i, f_0] = 0$ and $e_i |0\rangle = e_{i(0)} |0\rangle = 0$ for $i \geq 1$, and $e_0 v_m = 0$, by the representation theory of sl_2, since $h_0 v_m = m v_m$ and $e_0 |0\rangle = (e_{-\theta})_{(1)} |0\rangle = 0$.

Since the elements h_i span $\hat{\mathfrak{h}}'$ and the elements e_i generate the Lie algebra $\hat{\mathfrak{n}}_+$, it follows from (18.10) that $U(\hat{\mathfrak{g}}') v_m = U(\hat{\mathfrak{n}}_-) U(\hat{\mathfrak{h}}') U(\hat{\mathfrak{n}}_+) v_m = U(\hat{\mathfrak{n}}_-) v_m$. Since obviously the eigenvalues of the Sugawara operator $L_0^{\mathfrak{g}}$ on $U(\hat{\mathfrak{n}}_-) v_m$ do not exceed $-(m+1)$ and $L_0^{\mathfrak{g}} |0\rangle = 0$, it follows that $U(\hat{\mathfrak{g}}') v_m$ is a proper subrepresentation of $V^m(\mathfrak{g})$, hence is contained in $J^m(\mathfrak{g})$. This proves (18.9).

Let G be the connected simply connected algebraic group with Lie algebra $\mathfrak{g}$, and let $G e_\theta$ be the G-orbit of e_θ in $\mathfrak{g}$. By Proposition 17.2, the group G acts by automorphisms of the vertex algebra $V^m(\mathfrak{g})$. The group G leaves the ideal $J^m(\mathfrak{g})$ invariant, since it is the unique maximal ideal of $V^m(\mathfrak{g})$. It follows from (18.9) that

$$a_{(-1)}^{m+1} |0\rangle \in J^m(\mathfrak{g}) \quad \text{for any} \quad a \in G e_\theta. \tag{18.11}$$

Now it is easy to prove the following proposition.

Proposition 18.3. In any representation W of the vertex algebra $V_m(\mathfrak{g})$ one has:

$$a^W(z)^{m+1} = 0 \tag{18.12}$$

for every $a \in G e_\theta$.

Proof. Due to Proposition 18.2 and (18.11), we have:

$$Y^W(a_{(-1)}^{m+1} |0\rangle, z) = 0 \quad \text{if} \quad a \in G e_\theta.$$

But $Y(a_{(-1)} |0\rangle, z)$ is the current $a(z) = \sum_{n \in \mathbb{Z}} (a \otimes t^n) z^{-n-1}$ all of whose coefficients commute. Hence we have for $a \in G e_\theta$

$$
\begin{aligned}
0 &= Y^W(a_{(-1)}^{m+1} |0\rangle, z) \\
&= \; : Y^W(a_{(-1)}, z) \ldots Y^W(a_{(-1)}, z) : \quad (m+1 \text{ factors}) \\
&= \; : a^W(z) \ldots a^W(z) : \quad (m+1 \text{ factors}) \\
&= a^W(z)^{m+1}.
\end{aligned}
$$

∎

Given $N \in \mathbb{Z}$, the coefficient of z^{-N-m-1} of (18.12) is

$$\sum_{\substack{n_1, \ldots, n_{m+1} \in \mathbb{Z} \\ n_1 + \cdots + n_{m+1} = N}} a^W_{(n_1)} a^W_{(n_2)} \ldots a^W_{(n_{m+1})} = 0, \quad \text{if} \quad a \in G\, e_\theta. \quad (18.13)$$

Thus this complicated relation must hold in any representation W of the vertex algebra $V_m(\mathfrak{g})$. This is, of course, a very stringent condition on a representation W of the Kac-Moody affinization $\hat{\mathfrak{g}}'$, hence of the universal affine vertex algebra $V^m(\mathfrak{g})$, to be actually a representation of the simple affine vertex algebra $V_m(\mathfrak{g})$.

Let us see what this condition implies for $W = L(\lambda)$, the irreducible highest weight representation of $\hat{\mathfrak{g}}'$ with highest weight $\lambda \in \hat{\mathfrak{h}}'^*$ of level $\lambda(M) = m$. Recall that this is the unique (up to equivalence) irreducible representation of $\hat{\mathfrak{g}}'$, admitting a nonzero vector v_λ, such that (cf. Lecture 9):

$$\hat{\mathfrak{n}}_+ v_\lambda = 0, \quad h_i v_\lambda = \lambda(h_i) v_\lambda \quad \text{for} \quad i = 0, 1, \ldots, r.$$

Proposition 18.4. If the representation $W = L(\lambda)$ of $\hat{\mathfrak{g}}'$ can be extended to a representation of the vertex algebra $V_m(\mathfrak{g})$, where $m \in \mathbb{Z}_+$, then

$$\lambda(h_i) \in \mathbb{Z}_+ \quad \text{for all } i = 0, \ldots, r. \quad (18.14)$$

Proof. Since $a \otimes t^n \in \hat{\mathfrak{n}}_+$ for $a \in \mathfrak{g}$, $n > 0$, we have $a^W_{(n)} v_\lambda = 0$ for $a \in \mathfrak{g}$, $n > 0$. Hence identity (18.13) for $N = m + 1$ (resp. $N = 0$) gives:

$$(a^W_{(-1)})^{m+1} v_\lambda = 0 \quad \text{for} \quad a \in G\, e_\theta \quad (\text{resp. } (a^W_{(0)})^{m+1} v_\lambda = 0). \quad (18.15)$$

Putting $a = e_\theta$ in the first relation of (18.15) we get $f_0^{m+1} v_\lambda = 0$, hence, by the representation theory of sl_2, we get that $\lambda(h_0) \in \mathbb{Z}_+$. Since the Lie algebra $\mathfrak{g}$ is simple and the span of the orbit $G\, e_\theta$, being G-invariant, is an ideal of $\mathfrak{g}$, it follows that $\mathfrak{g}$ is spanned by this orbit. Hence there exists a basis $a_1, \ldots, a_d$ of $\mathfrak{g}$, consisting of elements from $G\, e_\theta$. Since all these elements are nilpotent, and, by the second relation in (18.15), $(a_i^W)^{m+1} v_\lambda = 0$ for all $i = 1, \ldots, d$, it follows by the PBW theorem that $\dim U(\mathfrak{g}) v_\lambda < \infty$. Hence $\lambda(h_i) \in \mathbb{Z}_+$ for all $i = 1, \ldots, r$. ∎

The highest weight representation $L(\lambda)$ of $\hat{\mathfrak{g}}'$ is called *integrable* if λ satisfies (18.14). These are precisely all the unitary highest weight representations of $\hat{\mathfrak{g}}'$ (cf. Lecture 9).

There are two ways to show that, conversely, any integrable representation $L(\lambda)$ of $\hat{\mathfrak{g}}'$ of level $m \in \mathbb{Z}_+$ extends to a representation of the vertex algebra $V_m(\mathfrak{g})$. The first way uses the isomorphism of the vertex algebra $V_1(\mathfrak{g})$ to the lattice vertex algebra V_Q, where Q is the root lattice of $\mathfrak{g}$, alluded to in Remark 17.4 (the so-called Frenkel-Kac construction, see Kac [1998]). This isomorphism allows one to construct all irreducible representations of $V_1(\mathfrak{g})$ explicitly, and then to use tensor products to construct all irreducible representations of $V_m(\mathfrak{g})$ for $m > 0$. The case $m = 0$ is trivial since $V_0(\mathfrak{g}) = \mathbb{C}|0\rangle$.

The second way uses the Zhu algebra, which is discussed in the next section.

18.5.   The Zhu algebra method

Let V be a conformal vertex algebra with the energy-momentum field $L(z) = \sum_n L_n z^{-n-2}$. A representation of V in a vector space M, for which L_0^M is diagonalizable with discrete real spectrum bounded below, is called a *positive-energy representation*; the eigenspace decomposition of M with respect to L_0^M can be written as follows:

$$M = \bigoplus_{j \geq h} M_j, \quad M_h \neq 0, \tag{18.16}$$

where M_j is eigenspace for L_0^M with eigenvalue j. For an eigenvector $a \in V$ with the eigenvalue Δ_a of the energy operator L_0 we write (cf. (17.23)):

$$Y^M(a, z) = \sum_{m \in -\Delta_a + \mathbb{Z}} a_m^M z^{-m-\Delta_a}.$$

Then the Borcherds identity (18.3) is rewritten for eigenvectors a and b of L_0 as follows ($k \in \mathbb{Z}$, $m \in -\Delta_a + \mathbb{Z}$, $n \in -\Delta_b + \mathbb{Z}$):

$$\sum_{j \in \mathbb{Z}_+} (-1)^j \binom{k}{j} \left(a_{m+k-j}^M b_{n+j}^M - (-1)^k p(a, b) b_{n+k-j}^M a_{m+j}^M \right)$$
$$= \sum_{j \in \mathbb{Z}_+} \binom{m + \Delta_a - 1}{j} \left(a_{(k+j)} b \right)_{m+n+k}^M. \tag{18.17}$$

In particular, for $k = 0$ we have the commutator formula (cf. (18.4))

$$[a_m^M, b_n^M] = \sum_{j \in \mathbb{Z}_+} \binom{m + \Delta_a - 1}{j} \left(a_{(j)} b \right)_{m+n}^M. \tag{18.18}$$

We also have an equivalent form of (18.7):

$$(L_{-1}a)_n^M = -(n + \Delta_a)a_n^M, \tag{18.19}$$

and a special case of (18.18):

$$[L_0^M, a_n^M] = -na_n^M. \tag{18.20}$$

Since $L_0 a = \Delta_a a$, we see from (18.19) that

$$((L_0 + L_{-1})a)_0^M = 0. \tag{18.21}$$

Also, by (18.20), vectors from $a_n^M M_h$ have L_0^M-eigenvalue $h - n$, hence we have:

$$a_n^M M_h = 0 \quad \text{if} \quad n > 0. \tag{18.22}$$

For the same reason $a_0^M M_h \subset M_h$, so that we have a linear map $\pi_M : V \to M_h$ defined by $\pi_M(a) = a_0|_{M_h}$.

From now on we assume that all eigenvalues of L_0^M are integers, and let $m = 1$, $k = -1$, $n = 0$ in the Borcherds identity (18.17). Then, due to (18.22), we obtain

$$a_0^M b_0^M v = (a * b)_0^M v, \quad v \in M_h, \tag{18.23}$$

where we define the product $*$ on V by the formula

$$a * b = \sum_{j \in \mathbb{Z}_+} \binom{\Delta_a}{j} (a_{(j-1)}b). \tag{18.24}$$

Formula (18.23) shows that the map π_M defines a representation of the algebra V with the (not necessarily associative) product $*$, in the space M_h (in the sense that $\pi_M(a * b) = \pi_M(a)\pi_M(b)$). Formula (18.21) shows that $\pi_M((L_{-1} + L_0)V) = 0$.

Consider the following subspace of V:

$$J = \text{span}_{\mathbb{C}} \{((L_{-1} + L_0)a) * b\}_{a,b \in V}.$$

Formulas (18.21) and (18.23) show that

$$\pi_M(J) = 0.$$

The above observations lead to the following remarkable result.

Theorem 18.2. (Zhu [1996])

(a) J is a two-sided ideal of the algebra $(V, *)$, which lies in the kernel of π_M.

(b) $Zhu\, V := (V, *)/J$ is a unital associative algebra.

(c) The map, which associates to a positive energy representation M of V the representation of the associative algebra $Zhu\, V$ in the vector space M_h, induces a map from equivalence classes of positive energy representations of V to equivalence classes of representations of the associative algebra $Zhu\, V$. This map is bijective on irreducible representations. ∎

A rather long proof of this theorem was given in Zhu [1996]. We refer the reader to a shorter and simpler proof in De Sole-Kac [2006], where also the assumption that L_0 has integer eigenvalues on V is removed (which requires a modification of the definition of the Zhu algebra $Zhu\, V$).

Example 18.2. Let $\mathfrak{g}$ be a simple Lie algebra with an invariant bilinear form $(\cdot|\cdot)$. It is easy to see that the linear map $\phi : U(\mathfrak{g}) \to Zhu\, V^m(\mathfrak{g})$, given by

$$a_1 \ldots a_k \mapsto\, : a_k \ldots a_1 : \mod J, \quad \text{where} \quad a_1, \ldots, a_k \in \mathfrak{g},$$

defines a surjective homomorphism of $U(\mathfrak{g})$ with the opposite multiplication to the Zhu algebra of the universal affine vertex algebra $V^m(\mathfrak{g})$. Given a representation of $\mathfrak{g}$ in a vector space U, we extend it to $\mathfrak{g}[t] \oplus \mathbb{C}K$, by letting $\mathfrak{g}[t]t$ act trivially and $M = m$, and consider the corresponding induced representation $\mathrm{Ind}_{\mathfrak{g}[t] \oplus \mathbb{C}M}^{\hat{\mathfrak{g}}} U$ of $\hat{\mathfrak{g}}$. This representation of $\hat{\mathfrak{g}}$ is obviously restricted, as is its quotient $L(U)$ by the maximal subrepresentation. By Theorem 18.1, $L(U)$ extends uniquely to a representation of $V^m(\mathfrak{g})$, and it is the irreducible representation of $V^m(\mathfrak{g})$ corresponding to the representation of $\mathfrak{g}$ in U. It follows that the homomorphism ϕ is injective, hence it is an isomorphism.

Example 18.3. Assume now that $\mathfrak{g}$ is simple with the invariant bilinear form $(\cdot|\cdot)$ normalized as in §18.4, and that $m \in \mathbb{Z}_+$. It is easy to deduce from the discussion in §18.4, that $Zhu\, V_m(\mathfrak{g}) \simeq U(\mathfrak{g})/(e_\theta^{m+1})$, and that this algebra is finite-dimensional. Hence all irreducible representations of the algebra $Zhu\, V_m(\mathfrak{g})$ are finite-dimensional with highest weight $\bar{\lambda}$, satisfying the condition $(\bar{\lambda}|\theta) \leq m$. The corresponding irreducible representations of the simple vertex algebra $V_m(\mathfrak{g})$ given by Theorem 18.2 (c) are precisely the integrable highest weight representations $L(\lambda)$, where $\lambda|_{\mathfrak{h}} = \bar{\lambda}$ and

$\lambda(M) = m$. For more details see Frenkel-Zhu [1992].

Example 18.4. As in Example 18.2, it is easy to see that the map $\phi :$ $\mathbb{C}[x] \to Zhu\, V^c$, given by

$$x^k \mapsto\; : \nu^k : \bmod J, \quad k \in \mathbb{Z}_+,$$

where ν is the conformal vector (17.24), defines an isomorphism of the algebra of polynomials $\mathbb{C}[x]$ and the algebra $Zhu\, V^c$. Moreover, the one-dimensional representation of $Zhu\, V^c$, given by $\nu \mapsto h \in \mathbb{C}$, corresponds to the irreducible representation $V(c, h)$ of the vertex algebra V^c.

Example 18.5. Consider now a simple Virasoro vertex algebra $V_c = V^c/J_c$ with $J_c \neq 0$ (then c is of the form described in Example 14.1). Its Zhu algebra is obviously a quotient of the algebra $Zhu\, V^c$ by a nonzero ideal, hence

$$Zhu\, V_c \simeq \mathbb{C}[x]/(P_c(x)),$$

where $P_c(x)$ is a monic polynomial of positive degree. It follows from Example 18.4 that the representation of V^c in $V(c, h)$ is actually a representation of V_c if and only if h is a root of the polynomial $P_c(x)$.

We explain here how to compute the polynomial $P_c(x)$ in the case $c = c_m$, given by (12.1 a). Namely, we shall show that all roots $h_{r,s}^{(m)}$ of $P_{c_m}(x)$ are given by (12.1 b), and they have multiplicity 1.

It follows from (12.17) for $r = s = 1$ that

$$\mathrm{ch}\, V_{c_m} \leq \frac{1}{\varphi(q)} \left(1 - q - q^{(m+1)(m+2)} + \text{higher degree terms} \right).$$

Since $\mathrm{ch}\, V^{c_m} = \frac{1-q}{\varphi(q)}$, we conclude that V^{c_m} has a singular vector v_m of energy e_m (i.e. $L_0 v_m = e_m v_m$, $L_n v_m = 0$ for $n > 0$), such that

$$e_m \leq (m + 1)(m + 2). \tag{18.25}$$

If an irreducible representation $V(c_m, h)$ of the vertex algebra V^{c_m} is actually a representation of the simple vertex algebra V_{c_m}, we necessarily have that

$$Y^{V(c_m, h)}(v_m, z) = 0. \tag{18.26}$$

It is not difficult to show (see Feigin-Fuchs [1984]) that in a PBW basis of V^{c_m}, described in §17.3, the coefficient of $L_{-2}^{(m+1)(m+2)/2}|0\rangle$ is nonzero, hence for a highest weight vector v of $V(c_m, h)$ we have:

$$z^{(m+1)(m+2)/2} Y^M(v_m, z) v|_{z=0} = Q(h)v,$$

where $Q(h)$ is a nonzero polynomial of degree $\frac{1}{2}(m+1)(m+2)$. Hence, by (18.26), $V(c_m, h_0)$ is a representation of V_{c_m} only when h_0 is a root of the polynomial $Q(h)$. It follows from (18.25) that

$$\text{(number of irreducible representations of } V_{c_m}) \leq \frac{1}{2}(m+1)(m+2). \quad (18.27)$$

On the other hand, identity (12.18) for $n = 0$ can be interpreted as:

$$V_1(sl_2) \otimes V_m(sl_2) = V_{m+1}(sl_2) \otimes V_{c_m}. \quad (18.28)$$

This follows from the fact that (12.16) is actually an equality (see Remark 12.3).

Furthermore, consider the following representation of the vertex algebra $V_1(sl_2) \otimes V_m(sl_2)$:

$$\mathcal{V} = \bigoplus_{j=0}^{m} (V_1(sl_2) \otimes L(j\omega_0 + (m-j)\omega_1)).$$

Using that (12.16) is an equality, formula (12.19) can be written as follows:

$$\mathcal{V}^{\hat{n}_+} = \bigoplus_{\substack{r,s \in \mathbb{N}, \\ 1 \leq s \leq r \leq m+1}} L(c_m, h_{r,s}^{(m)}),$$

and interpreted as a decomposition of the representation of the vertex algebra V_{c_m} in $\mathcal{V}^{\hat{n}_+}$. We thus have constructed $\frac{1}{2}(m+1)(m+2)$ inequivalent irreducible representations of V_{c_m}. Due to inequality (18.27), these are all irreducible representations of V_{c_m}. At the same time, we computed the polynomial $P_{c_m}(x)$.

Using more general, modular invariant representations, studied in Kac-Wakimoto [1988], one can derive a similar result for an arbitrary c, given formula (14.6), i.e. treat all cases when V^c is not simple. The result was first obtained by a different method in Wang [1993].

Remark 18.1. A vertex algebra is called *semisimple* if any of its representations is completely reducible. It follows from Theorem 16.2 that the vertex algebra $F(A)$ of free fermions is semisimple, provided that A is a purely odd superspace. It is known that the vertex algebra $V_m(\mathfrak{g})$ is semisimple if and only if $m \in \mathbb{Z}_+$, and that $V_c(m)$ is semisimple if and only if c is of the form (14.6) (see Dong-Li-Mason [1997]). ■

18.6. Twisted representations

In order to explain the idea of a twisted representation of a vertex algebra (which has no analogue in the representation theory of associative or Lie algebras), consider the following two examples.

Example 18.6. Let σ be a finite order automorphism of a Lie algebra $\mathfrak{g}$, so that $\sigma^N = 1$ for some positive integer N, and let $\Gamma = \frac{1}{N}\mathbb{Z}$. Then we have a $\Gamma/\mathbb{Z}$-grading

$$\mathfrak{g} = \bigoplus_{\bar\alpha \in \Gamma/\mathbb{Z}} \mathfrak{g}_{\bar\alpha},$$

where $\mathfrak{g}_{\bar\alpha}$ is the eigenspace of σ attached to the eigenvalue $\exp(2\pi i\bar\alpha)$. Consider the loop algebra $\tilde{\mathfrak{g}} = \mathfrak{g}[t^{\frac{1}{N}}, t^{-\frac{1}{N}}]$ (in Example 13.2 $\tilde{\mathfrak{g}} = \mathfrak{g}[t, t^{-1}]$, but, replacing t by $t^{1/N}$, we get an isomorphic Lie algebra). Define the σ-*twisted loop algebra* as the following subalgebra $\tilde{\mathfrak{g}}_\sigma$ of $\tilde{\mathfrak{g}}$:

$$\tilde{\mathfrak{g}}_\sigma = \bigoplus_{\alpha \in \Gamma} \mathfrak{g}_{\bar\alpha} t^\alpha.$$

Here and further $\bar\alpha$ denotes the coset $\alpha + \mathbb{Z}$ in the group Γ. If $\alpha, \beta \in \Gamma$, letting $a_{(\alpha)} = at^\alpha$, $b_{(\beta)} = bt^\beta$, where $a \in \mathfrak{g}_{\bar\alpha}$, $b \in \mathfrak{g}_{\bar\beta}$, we get

$$[a_{(\alpha)}, b_{(\beta)}] = [a, b]_{(\alpha+\beta)}$$

with $[a, b] \in \mathfrak{g}_{\bar\alpha+\bar\beta}$. The *twisted affine Kac-Moody algebra* (cf. Kac [1983]) is the Lie algebra $\hat{\mathfrak{g}}'_\sigma = \tilde{\mathfrak{g}}_\sigma \oplus \mathbb{C}M$ with the bracket

$$[a_{(\alpha)}, b_{(\beta)}] = [a, b]_{(\alpha+\beta)} + \alpha\delta_{\alpha, -\beta}(a|b)M, \quad \alpha, \beta \in \Gamma, a \in \mathfrak{g}_{\bar\alpha}, b \in \mathfrak{g}_{\bar\beta}.$$

(This is the usual Kac-Moody affinization if $N = 1$.) Letting

$$a^{tw}(z) = \sum_{m \in \bar\alpha} a_{(m)} z^{-m-1}, \quad b^{tw}(z) = \sum_{n \in \bar\beta} b_{(n)} z^{-n-1},$$

we get an equivalent form of the bracket in $\hat{\mathfrak{g}}'_\sigma$:

$$[a^{tw}(z), b^{tw}(w)] = [a, b]^{tw}(w)\delta_{\bar\alpha}(z, w) + (a|b)\partial_w \delta_{\bar\alpha}(z, w)M, \tag{18.29}$$

where

$$\delta_{\bar{\alpha}}(z, w) = \left(\frac{w}{z}\right)^{\alpha} \delta(z, w)$$

is the "twisted" formal delta function (which depends only on the coset $\alpha + \mathbb{Z}$). So we still have the locality property for a pair of "twisted" formal distributions $(a(z), b(z))$ (the locality is defined in the same way as in Lecture 13 for ordinary formal distributions).

Example 18.7. In §15.3 we studied the neutral free fermion $\psi(z) = \sum_{n \in \mathbb{Z}} \psi_{(n)} z^{-n-1} = \sum_{n \in \frac{1}{2} + \mathbb{Z}} \psi_n z^{-n-1/2}$, which is an odd formal distribution, and the commutation relations are

$$[\psi_{(m)}, \psi_{(n)}] = \delta_{m, -n-1}, \, m, n \in \mathbb{Z}$$

$$\left(\text{respectively } [\psi_m, \psi_n] = \delta_{m, -n}, \, m, n \in \frac{1}{2} + \mathbb{Z}\right),$$

or, equivalently, $[\psi(z), \psi(w)] = \delta(z, w)$. Consider the *twisted neutral fermion*

$$\psi^{tw}(z) = \sum_{n \in \mathbb{Z}} \psi_n^{tw} z^{-n-1/2} = \sum_{n \in \frac{1}{2} + \mathbb{Z}} \psi_{(n)}^{tw} z^{-n-1}$$

with commutation relations

$$[\psi_m^{tw}, \psi_n^{tw}] = \delta_{m, -n}, \, m, n \in \mathbb{Z}$$

$$\left(\text{respectively } [\psi_{(m+\frac{1}{2})}^{tw}, \psi_{(n+\frac{1}{2})}^{tw}] = \delta_{m, -n-1}, \, m, n \in \frac{1}{2} + \mathbb{Z}\right),$$

which is equivalent to

$$[\psi^{tw}(z), \psi^{tw}(w)] = \delta_{1/2}(z, w). \tag{18.30}$$

Definition 18.2. Let V be a vertex algebra and let σ be a finite order automorphism of V, so that $\sigma^N = 1$ for some positive integer N. Let $V = \bigoplus_{\bar{\alpha} \in \Gamma / \mathbb{Z}} V^{\bar{\alpha}}$, where $V^{\bar{\alpha}}$ is the eigenspace of σ, attached to the eigenvalue $\exp(2\pi i \bar{\alpha})$. A σ-*twisted representation* of the vertex algebra V in a vector superspace M is a parity preserving map

$$a \mapsto Y^M(a, z), \quad Y^M(a, z) = \sum_{n \in \bar{\alpha}} a_{(n)}^M z^{-n-1}, \, a \in V^{\bar{\alpha}},$$

where $a_{(n)}^M \in \text{End } M$, with the property that for each $v \in M$, $a_{(n)}^M v = 0$ if n is sufficiently large, and such that (18.1) and the *twisted Borcherds identity* hold. The latter is obtained from (18.2) by replacing $\delta(z, w)$ by $\delta_{\bar{\alpha}}(z, w)$.

Note that picking $m \in \bar{\alpha}$ and $n \in \bar{\beta}$ and equating the coefficients of $z^{-m-1}w^{-n-1}$ in the twisted Borcherds identity, we obtain exactly equation (18.3) (except that m and n are not integers anymore).

Letting $k = 0$ in the twisted Borcherds identity, we obtain the twisted commutator formula (which looks exactly as (17.21)):

$$[Y^M(a,z), Y^M(b,w)] = \sum_{j \in \mathbb{Z}_+} Y^M(a_{(j)}b, w) \partial_w^j \delta_{\bar{\alpha}}(z,w)/j!. \qquad (18.31)$$

Note that the automorphism σ of $\mathfrak{g}$ in Example 18.6 extends to an automorphism of the affine algebra $\hat{\mathfrak{g}}'$, by letting $\sigma(at^\alpha) = \sigma(a)t^\alpha$, $\sigma(M) = M$. The latter induces an automorphism of the universal affine vertex algebra $V^m(\mathfrak{g})$, again denoted by σ, such that $\sigma^N = 1$. Comparing (18.29) with (18.31), we see that any σ-twisted representation of the vertex algebra $V^m(\mathfrak{g})$ is automatically a restricted representation of level m of the twisted affine Kac-Moody algebra $\hat{\mathfrak{g}}'_\sigma$.

Likewise, the map $\psi_n \mapsto -\psi_n$ induces an order two automorphism σ of the neutral free fermion vertex algebra, and, comparing (18.30) with (18.31), we see that a σ-twisted representation of this vertex algebra is a restricted representation of the Lie superalgebra (18.30).

In the same way as in the non-twisted case, we show that (18.7) holds also for any twisted representation M of V.

Finally, let $k = -1$ in the twisted Borcherds identity. Multiply both sides by $(z/w)^\alpha$ for some α in the coset $\bar{\alpha} \in \Gamma/\mathbb{Z}$, and take Res_z of both sides. It is easy to see that we obtain the following twisted version of the (-1)st product identity (18.6) (after replacing w by z):

$$: Y^M(a,z)Y^M(b,z) := \sum_{j=0}^{\infty} \binom{\alpha}{j} Y^M(a_{(j-1)}b, z)z^{-j}, \qquad (18.32)$$

where the normal ordered product is defined by the same formula as (14.11) with the negative and positive parts

$$Y^M(a,z)_- = \sum_{j \in \alpha + \mathbb{Z}_+} a_{(j)}^M z^{-j-1}, \quad Y^M(a,z)_+ = Y^M(a,z) - Y^M(a,z)_-.$$

Note that both sides of (18.32) depend on the choice of α, and that this identity coincides with (18.6) if $\alpha = 0$.

The following theorem is a twisted analogue of Theorem 18.1. (Both theorems hold for any universal enveloping algebra of a Lie conformal algebra $\mathcal{R}$, see Remark 17.5, and any automorphism of $\mathcal{R}$, but we state them only for the examples at hand.)

Theorem 18.3. Any restricted representation W of level m of the twisted affine algebra $\hat{\mathfrak{g}}'_\sigma$ (resp. of the Lie superalgebra (18.30)) uniquely extends to a σ-twisted representation of the universal affine vertex algebra $V^m(\mathfrak{g})$ (resp. of the neutral free fermion vertex algebra).

Proof. The uniqueness is clear by formulas (18.31), (18.32), (18.7). The proof of existence is based on a twisted version of the "abstract" example, given by Proposition 17.1. Namely, we define the n-th product ($n \in \mathbb{Z}$) of twisted local continuous formal distributions by Bakalov's formula (14.16), i.e.

$$a(w)_{(n)}b(w) = (\partial_z^{N-n-1}F(z,w)/(N-n-1)!))|_{z=w},$$

verify that the vacuum, translation covariance and locality axioms hold, thus we obtain a vertex algebra. Its tautological twisted representation $a \mapsto a(z)$ in W lifts to the twisted representation of the vertex algebra in question, by its universality property. A proof of existence, based on a more complicated definition of the n-th products, can be found in Li [1996]. ∎

Example 18.8. Consider the representation of the Lie superalgebra (18.30) in the vector superspace $M = V_{\delta=0}$, constructed in §3.4. By Theorem 18.3, it extends to a σ-twisted representation of the neutral free fermion vertex algebra F. Let $\psi = \psi_{(-1)}|0\rangle \in F$. Then $\psi(z) = Y(\psi, z)$ and $\psi^{tw}(z) = Y^M(\psi, z)$. To simplify notation, we let $a^{tw}(z) = Y^M(a, z)$ for arbitrary $a \in F$. Recall that $[\psi_\lambda \psi] = 1$, and that in §15.3 we constructed the state $L = \frac{1}{2} : T\psi\psi :$, such that the corresponding quantum field is the Virasoro field $L(z) = \frac{1}{2} : \partial\psi(z)\psi(z) :$ with central charge $c = \frac{1}{2}$. Let $L^{tw}(z) = Y^M(L, z)$. Since $\partial\psi_{(j)}\psi = 0$ for $j \neq 1$ and $\partial\psi_{(1)}\psi = -1$, we obtain by formula (18.32) with $\alpha = 1/2$:

$$\frac{1}{2} : \partial\psi^{tw}(z)\psi^{tw}(z) := L^{tw}(z) + \frac{1}{2}(\partial\psi_{(0)}\psi)^{tw}(z)z^{-1}$$
$$+ \frac{1}{2}\binom{1/2}{2}(\partial\psi_{(1)}\psi)^{tw}(z)z^{-2} + \dots.$$

Hence

$$L^{tw}(z) = \frac{1}{2} : \partial\psi^{tw}(z)\psi^{tw}(z) : -\frac{1}{16z^2}.$$

Applying (18.31) to $a = b = L$, hence $\bar{\alpha} = 0$, we obtain that $L^{tw}(z) = \sum_{j \in \mathbb{Z}} L_j^{tw} z^{-j-2}$ is a Virasoro field with central charge $c = 1/2$. In particular, we have

$$L_0^{tw} = \sum_{n \geq 1} n \psi_{-n}^{tw} \psi_n^{tw} + \frac{1}{16}.$$

This is consistent with the construction of the Ramond sector in §3.4.

Example 18.9. Let σ be a finite order automorphism of a simple Lie algebra $\mathfrak{g}$. Consider a restricted representation W of level m of the twisted affine algebra $\hat{\mathfrak{g}}_\sigma'$. By Theorem 18.3, it extends to a σ-twisted representation of the universal affine vertex algebra $V^m(\mathfrak{g})$. Let $\{u_i\}$, $\{u^i\}$ $(i = 1, \ldots, \dim \mathfrak{g})$ be dual bases of $\mathfrak{g}$ with respect to a nondegenerate invariant bilinear form, compatible with the eigenspace decomposition for σ. Recall that $V^m(\mathfrak{g})$ is a conformal vertex algebra with a Virasoro field $L^{\mathfrak{g}}(z)$ defined in §16.2. Similarly to Example 18.8, denote $a^{tw}(z) = Y^W(a, z)$ for arbitrary $a \in V^m(\mathfrak{g})$. Then $u_i^{tw}(z) = Y^W(u_i, z)$, $u^{i,tw}(z) = Y^W(u^i, z)$, $L^{\mathfrak{g},tw}(z) = Y^W(L^{\mathfrak{g}}, z)$, and by (18.32),

$$: u_i^{tw}(z) u^{i,tw}(z) : = \sum_{j=0}^{\infty} \binom{\alpha_i}{j} Y^W(u_{i\,(j-1)} u^i, z) z^{-j},$$

where α_i is a representative of the coset $\bar{\alpha}_i \in \Gamma/\mathbb{Z}$ such that $u_i \in \mathfrak{g}_{\bar{\alpha}_i}$. Using the λ-bracket for currents (13.31), we get that $Y^W(u_{i\,(j-1)} u^i, z) = 0$ for $j > 2$, and $Y^W(u_{i\,(-1)} u^i, z) = 2(m+g) L^{\mathfrak{g},tw}(z)$, $Y^W(u_{i\,(0)} u^i, z) = [u_i, u^i]^{tw}(z)$, $Y^W(u_{i\,(1)} u^i, z) = m$, so

$$L^{\mathfrak{g},tw}(z) = \frac{1}{2(m+g)} \left(\sum_{i=1}^{\dim \mathfrak{g}} : u_i^{tw}(z) u^{i,tw}(z) : \right.$$
$$\left. - \sum_{i=1}^{\dim \mathfrak{g}} \alpha_i [u_i, u^i]^{tw}(z) \frac{1}{z} - \sum_{i=1}^{\dim \mathfrak{g}} \frac{\alpha_i(\alpha_i - 1)}{2} \frac{m}{z^2} \right).$$

By commutator formula (18.31) and relations (16.10), $L^{\mathfrak{g},tw}(z)$ is a Virasoro field with central charge $c(m) = \frac{m \dim \mathfrak{g}}{m+g}$. A direct proof of this fact, as for the ordinary Sugawara construction in Lecture 10, is quite involved.

REFERENCES

Arbarello, E., De Concini, C., Kac, V.G., Procesi, C. [1988] "Moduli spaces of curves and representation theory", *Comm. Math. Phys.*, **117** (1988), 1–36.

Astashkevich, A. [1997] "On the structure of Verma modules over Virasoro and Neveu-Schwarz algebras". *Comm. Math. Phys.*,**186** (1997), no. 3, 531–562.

Bakalov, B., Kac, V. G. [2003] "Field algebras", *IMRN* no. 3 (2003), 123–159.

Bararkat, A., De Sole, A., Kac, V.G. [2009] "Poisson vertex algebras in the theory of Hamiltonian equations", *Jpn. J. Math.* **4** (2009), no. 2, 141–252.

Belavin, A.A., Polyakov, A.M., Zamolodchikov, A.B. [1984a] "Infinite conformal symmetry in two-dimensional quantum field theory", *Nuclear Phys.* **B241** (1984), 333–380.

Belavin, A.A., Polyakov, A.M., Zamolodchikov, A.B. [1984b] "Infinite conformal symmetry of critical fluctuations in two dimensions", *J. Statist. Phys.* **34** (1984), 763–774.

Benkart, G. [1986] A Kac-Moody bibliography and some related references, in: *Lie algebras and related topics*, CMS conference proceedings 5, 1986, 111–138.

Borcherds R. [1986], "Vertex algebras, Kac-Moody algebras, and the Monster", *Proc. Nat. Acad. Sci. U.S.A.* **83** (1986), no. 10, 3068–3071.

Capelli, A., ltzykson, C., Zuber, J. B. [1987] "The A-D-E classification of minimal and $A_1^{(1)}$ conformal invariant theories", *Comm. Math. Phys.* **113**

(1987) no. 1, 1–26.

Chari, V., Pressley, A. [1987] "Unitary representations of the Virasoro algebra and a conjecture of Kac", *Compositio Math.* **67** (1988), no. 3, 315–342.

Chodos, A., Thorn, C.B. [1974] "Making the massless string massive", *Nuclear Phys.* **B72** (1974), 509–522.

Dashen, R., Frishman, Y. [1975] "Four-fermion interactions and scale invariance", *Phys. Rev.* **D11** (1975), 2781–2802.

Date, E., Jimbo, M., Kashiwara, M., Miwa, T. [1981] "Operator approach to the Kadomtsev-Petviashvili equation. Transformation groups for soliton equations III", *J. Phys. Soc. Japan* **50** (1981), 3806–3812.

Date, E., Jimbo, M., Kashiwara, M., Miwa, T. [1983] "Transformation groups for soliton equations", in: *Proceedings of RIMS Symposium*, M. Jimbo and T. Miwa, eds., World Scientific, 1983, 39–120.

De Sole, A., Kac, V.G. [2006] "Finite vs affine W-algebras", *Japanese Journal of Mathematics* **1** (2006), no. 1, 137–261.

Dirac, P.A.M. [1958] *The Principles of Quantum Mechanics*, 4th edition, Oxford University Press, 1958.

Dong, C., Li H., Mason, G. [1997] "Regularity of rational vertex operator algebras", *Advances in Math* **132** (1997), 148–166.

Feigin, B.L. [1984] "Semi-infinite homology of the Kac-Moody and Virasoro algebras", *Uspekhi Mat. Nauk* **39** (1984), 195–196. English translation: *Russian Math. Surveys* **39** (1984), No. 2, 155–156.

Feigin, B.L., Fuchs, D.B. [1982] "Invariant skew-symmetric differential operators on the line and Verma modules over the Virasoro algebra", *Funkt. Anal. i ego Prilozh.* **16** (1982), No. 2, 47–63, English translation: *Funct. Anal. Appl.* **16** (1982), 114–126.

Feigin, B.L., Fuchs, D.B. [1983a] "Verma modules over the Virasoro algebra", *Funkt. Anal. i ego Prilozh.* **17** (1983), No. 3, 91–92. English translation: *Funct. Anal. Appl.* **17** (1983), 241–242.

Feigin, B.L., Fuchs, D.B. [1983b] "Representations of the Virasoro algebra", Moscow University preprint.

Feigin, B.L., Fuchs, D.B. [1984] "Verma modules over the Virasoro algebra", in *Lecture Notes in Math.* **1060** (1984), 230–245.

Feingold, A.J., Lepowsky, J. [1978] "The Weyl-Kac character formula and power series identities", *Adv. in Math.* **29** (1978), 271–309.

Frenkel, I.B. [1981] "Two constructions of affine Lie algebra representations and boson-fermion correspondence in quantum field theory", *J. Funct. Anal.* **44** (1981), 259–327.

Frenkel, I.B., Garland, H., Zuckerman G. [1986], "Semi-infinite cohomology and string theory", *Proc. Nat. Acad. Sci. USA,* **83** (1986), 8442–8446.

Frenkel, I.B., Zhu, Y. [1992] "Vertex operator algebras associated to representations of affine and Virasoro algebras", *Duke Math. J.* **66** (1992), 123–168.

Friedan, D., Qiu, Z., Shenker, S. [1985] "Conformal invariance, unitarity and two-dimensional critical exponents", in *Publ. MSRI No.* **3** (1985), 419–449.

Friedan, D., Qiu Z., Shenker, S. [1986] "Details of the non-unitarity proof for highest weight representations of the Virasoro algebra", *Comm. Math. Phys.* **107** (1986), 535–542.

Fuchs, D.B. [1984] *Cohomology of infinite dimensional Lie algebras*, Nauka, Moscow, 1984; English Translation: Plenum Press, New York, 1986.

Garland, H. [1978] "The arithmetic theory of loop algebras", *J. Algebra* **53** (1978), 480–551.

Gelfand, I.M., Fuchs, D.B. [1968] "The cohomology of the Lie algebra of vector fields in a circle", *Funkt. Anal. i ego Prilozh.* **2** (1968), No. 4, 92–93 English translation: *Functional Anal. Appl.,* **2** (1968), 342–343.

Gepner, D., Witten, E. [1986] "String theory on group manifold", *Nucl. Phys.* **B278** (1986), 493–549

Goddard, P., Kent, A., Olive, D. [1985] "Virasoro algebra and coset space models", *Phys. Lett.* **B152** (1985), 88–92.

Goddard, P., Kent. A., Olive, D. [1986] "Unitary representations of the Virasoro and super-Virasoro algebras", *Comm. Math. Physics* **103** (1986), 105–119.

Goddard, P., Olive, D. [1986] "Kac-Moody and Virasoro algebras in relation to quantum physics", *Internat. J. Mod. Phys.* **A1** (1986), 303–414.

Goodman, R., Wallach, N. [1985] "Projective unitary positive-energy representations of Diff(S^1)", *J. Funct. Anal.* **63** (1985), 299–321.

Gorelik, M., Kac, V.G. [2007] "On simplicity of vacuum modules", *Adv. Math.* **211** (2007), 621 – 677.

Jimbo, M., Miwa, T. [1983] "Solitons and infmite dimensional Lie algebras", *Publ. Res. Inst. Math. Sci.* **19** (1983), 943–1001.

Kac, V.G. [1974] "Infinite dimensional Lie algebras and Dedekind's η-function", *Funkt. Anal. i ego Prilozh.* **8** (1974), No. 1, 77–78. English translation: *Functional Anal. Appl.* **8** (1974), 68–70.

Kac, V.G. [1978] "Highest weight representations of infinite dimensional Lie algebras", in: *Proceedings of ICM*, 299–304, Helsinki, 1978.

Kac, V.G. [1979] "Contravariant form for the infinite-dimensional Lie algebras and superalgebras", in: *Lecture Notes in Physics* **94** (1979) 441–445.

Kac, V.G. [1982] "Some problems on infinite-dimensional Lie algebras and their representations", in: *Lecture Notes in Math.* **933** (1982), 117–126.

Kac, V.G. [1983] "Infinite dimensional Lie algebras", *Progress in Mathematics* **44**, Birkhäuser, Boston, 1983. Second edition, Cambridge University Press, 1985. Third Edition, Cambridge University Press, 1990.

Kac, V. G. [1998] "Vertex Algebras for Beginners", second edition, University Lecture Series, Vol.10, American Mathematical Society, Providence, Rhode Island, 1998.

Kac, V.G., Kazhdan, D.A. [1979] "Structure of representations with highest weight of infinite-dimensional Lie algebras", *Adv. in Math.* **34** (1979), 97–108.

Kac, V.G., Kazhdan, D.A., Lepowsky J., Wilson, R.L. [1981] "Realization of the basic representations of the Euclidean Lie algebras", *Adv. in Math.* **42** (1981), 83–112.

Kac, V.G., Peterson, D.H. [1981] "Spin and wedge representations of infinite-dimensional Lie algebras and groups", *Proc. Nat. Acad. Sci. USA* **78** (1981), 3308–3312.

Kac, V.G., Peterson, D.H. [1984a] "Infinite dimensional Lie algebras, theta functions and modular forms", *Adv. in Math.* **53** (1984), 125–264.

Kac, V.G., Peterson, D.H. [1984b] "Unitary structure in representations of infinite dimensional groups and a convexity theorem", *Invent. Math.* **76** (1984), 1–14.

Kac, V.G., Peterson, D.H. [1986] "Lectures on the infinite wedge representation and the MKP hierarchy", in: *Séminaire de Math. Sup.* **102**, Montreal University, (1986), 141–184.

Kac, V. G., van de Leur, J. W. [1993] "The n-component KP hierarchy and representation theory. Important developments in soliton theory", in: *Springer Ser. Nonlinear Dynam.*, (1993), 302–343.

Kac, V.G., Wakimoto, M. [1986] "Unitarizable highest weight representations of the Virasoro, Neveu-Schwarz and Ramond algebras", in: *Lecture Notes in Physics*, **261** (1986), 345–371.

Kac, V.G., Wakimoto, M. [1987] "Modular and conformal invariance constraints in representation theory of affine algebras", *Adv. in Math.*, **70** (1988), no. 2, 156–236.

Kac, V. G., Wakimoto, M., [1988] "Modular invariant representations of infinite-dimensional Lie algebras and superalgebras", *Proc. Natl. Acad. Sci. USA* **85** (1988), 4956–4960.

Kaplansky, I., Santharoubane, L.J. [1985] "Harish-Chandra modules over the Virasoro algebra", in: *MSRI Publ.* **4** (1985), 217–231.

Kashiwara, M., Miwa, T. [1981] "The τ-function of the Kadomtsev-Petviashivili equation", *Proc. Japan Acad.* **57A** (1981), 342–347.

Kent, A. [1991] "Signature characters for the Virasoro algebra", *Phys. Lett.* B **269** (1991), 315–318.

Langlands, R. [1986] "On unitary representations of the Virasoro algebra", *Infinite-dimensional Lie algebras and their applications* (Montreal, PQ, 1986), World Sci. Publ., 1988, 141–159.

Li, H. [1996] "Local systems of twisted vertex operators, vertex operator superalgebras and twisted modules", *Contemp. Math.* **193** (1996), 203–236.

Macdonald, I.G. [1979] *Symmetric functions and Hall polynomials*, Oxford University Press, 1979.

Mathieu, O. [1992] "Classification of Harish-Chandra modules over the Virasoro algebra", *Invent. Math.* **107** (1992), 225–234.

Sato, M. [1981] "Soliton equations as dynamical systems on infinite dimensional Grassmann manifolds", *RIMS Kokyoroku* **439** (1981), 30–46.

Segal, G. [1981] "Unitary representations of some infinite-dimensional groups", *Comm. Math. Phys.* **80** (1981), 301–342.

Skyrme, T.H.R. [1971] "Kinks and the Dirac equation," *J. Math. Phys.* **12** (1971), 1735–1743.

Sugawara, H. [1968] "A field theory of currents", *Phys. Rev.* **170** (1968), 1659–1662.

Thorn, C.B. [1984] "Computing the Kac determinant using dual model techniques and more about the no-ghost theorem", *Nuclear Phys.* **B248** (1984), 551–569.

Tsuchiya, A., Kanie, Y. [1986] "Unitary representations of the Virasoro algebra", *Duke Math. J.* **53** (1986), 1013–1046.

Virasoro, M.A. [1970] "Subsidiary conditions and ghosts in dual-resonance models", *Phys. Rev.* **D1** (1970), 2933–2936.

Wakimoto, M., Yamada, H. [1986] "The Fock representations of the Virasoro algebra and the Hirota equations of the modified KP hierarchy", *Hiroshima Math. J.* **16** (1986), 427–441.

Wang, W. [1993] "Rationality of Virasoro vertex operator algebras", *IMRN* **71** (1993), 197–211.

Wightman, A.S. [1956] "Quantum field theory in terms of vacuum expectation values", *Phys. Rev.* **101** (1956), 860–866.

Zhu, Y. [1996] "Modular invariance of characters of vertex operator algebras", *J. AMS* **9** (1996), 237–301.

For other references see Kac [1983], Goddard-Olive [1986], and Benkart [1986].

INDEX

affine algebra, 93
affine Kac-Moody algebra, 93, 94
  twisted, 223
annihilation operator, 13
anticommutator, 48

Borcherds identity, 208, 212
  twisted, 224
boson-fermion correspondence, 46,
  183, 200
bosonic Fock representation, 157, 183
bosonization, 179

Cartan subalgebra, 97
Casimir operator, 102, 124, 178
central charge, 15, 43, 155, 170, 174,
  179, 204
centralizer, 204
character, 22, 115
charge, 36, 182
charged free fermions, 172, 200
Clifford affinization, 149, 157, 170,
  171, 187
Clifford algebra, 48
complete symmetric functions, 56
completion, 153
conformal vector, 204
conformal weight, 175
contracting operator, 48, 184
coset models, 204
creation operator, 13
current, 147, 158

cutoff procedure, 16, 104

Decomposition Theorem, 140
degree, 14, 34
derivation, 203
  inner, 204
Dong's lemma, 167
dual Coxeter number, 106, 178

electron, 33
energy, 34, 182
energy operator, 5, 11, 19, 43, 204
energy-momentum field, 204
Euler identity, 119
Existence Theorem, 195
expectation value, 25

fermionic Fock representation, 42,
  157, 182, 188
Fock space, 12
form
  contravariant, 4
  invariant, 93
Formal Cauchy formulas, 158
formal delta function, 137
  twisted, 224
formal distribution, 135
  continuous, 154
  primary, 175
formal Fourier transform, 143
free boson, 147, 156, 158, 180, 198,
  205